The Quantum Leap:
Next Generation

The Manufacturing Strategy For Business

Based on the original work
by **John R. Costanza**

As updated by
Dean Gilliam
& Steve Taylor-Jones

Foreword by **Carol Ptak**
CFPIM, CIRM, Jonah, PMP

Copyright ©2005 by JCIT International

ISBN 1-932159-44-4

Printed and bound in the U.S.A. Printed on acid-free paper
10 9 8 7 6 5 4 3 2

Library of Congress Cataloging-in-Publication Data
Gilliam, Dean, 1953-
Quantum leap : the next generation / by Dean Gilliam and Steve Taylor Jones.
p. cm.
Includes index.
ISBN 1-932159-44-4 (hardcover : alk. paper)
1. Production management. 2. Production control. 3. Just-in-time systems. I.
Jones, Steve Taylor. II. Title.
TS155.G52 2004
658.5--dc22

                            2004021792

Direct all inquiries to
J. Ross Publishing, Inc., 6501 Park of Commerce Blvd.
Suite 200, Boca Raton, Florida 33487.
Phone: (561) 869-3900
Fax: (561) 892-0700
Web: www.jrosspub.com

Table of Contents

Quantum Leap
The Next Generation

Foreword
By Carol A. Ptak, CFPIM, CIRM, Jonah, PMP

Historical Perspective

The world of manufacturing today bears little resemblance to the world of 30 years ago when I first walked onto a shop floor as an assembler. In those days, labor made up at least one-third of the cost of goods; inventory was considered an asset in both accounting and real terms; and time was measured in weeks and months. APICS, the Educational Society for Resource Management, was conducting a material requirements planning (MRP) crusade. The thought leaders of the day were Joe Orlicky, George Plossl, and Ollie Wight. In response to market forces and competitive requirements, their visionary leadership gave birth to an entirely new industry of software to manage the inventory necessary to achieve the high labor efficiencies that defined success at the time.

High-performing companies shared performance standards and key indicators to help other companies follow the proven path to success. Surprisingly, items such as input data accuracy continue to pose significant challenges to today's manufacturing companies, potentially inhibiting the implementation of much more sophisticated and powerful software. The late '70s and early '80s saw the emergence of the concept of just-in-time (JIT), with great successes at early adopters such as Hewlett Packard. Quickly, manufacturing costs began to shift from labor to material as manufacturers focused on their core competencies. A few lone voices in the wilderness advocated a different vision of manufacturing. Dick Ling developed and evangelized the idea of sales and operations planning — an idea only now receiving support from commercial software.

Dr. W. Edwards Deming began his quality crusade work in the United States after an amazing success transforming the meaning of "Made in Japan" from cheap, poor-quality goods to the "Lexus-quality" standard that all manufacturers strive for today. His work is the foundation behind the popular Six Sigma improvement concept. Dr. Eliyahu Goldratt shocked the world with his business book that was a novel (or was that a novel that was also a business book?). In either case, it taught the lesson that a goal, and the constraints to achieve that goal, must be identified and managed. How many forget this common sense and suffer for it?

John Costanza earned a Nobel Prize nomination for his early work in Demand Flow manufacturing. The first edition of *Quantum Leap* significantly challenged the status quo in manufacturing, with ideas that contradicted the widely accepted notions of the time. Who could forget his early evangelists at APICS conferences with their "MRP Not" buttons? However, the true measure of success is not what the passionate visionary achieves but rather the passion and legacy that others pick up and extend. As CEO and COO of JCIT, Dean Gilliam and Tony Gorski burn with the same passion and fire about the world of manufacturing. This was apparent in our first meeting, and a close friendship developed quickly. Rather than embracing a traditional consulting approach, where the consultant becomes an integral part of the company and revenue accrues in perpetuity to the consulting company, Dean and Tony embrace the concept of "teaching their customers to fish so they can eat for a lifetime." They have educated more than 90,000 people in the concepts of Demand Flow Technology. With their pragmatic approach and an understanding of the synergy between technology and their breakthrough ideas, they now have brought the concepts of Demand Flow Technology manufacturing to the masses around the world.

This second edition of *Quantum Leap* attempts to capture the knowledge and learning that has been accrued since its initial publication in 1990. When Tony and Dean asked me to write the foreword and introduction, I was deeply flattered. As an author of books on MRP, enterprise resource planning (ERP), and theory of constraints (TOC), I was also struck by the convergence of ideas in the past few years.

How to Use this Book

This book is not intended to be read from cover to cover, as you would read a novel. Nor is a simple browsing sufficient. This is a book that you will use often as you start the transformation process to demand-driven manufacturing. If you are the senior executive, then the introduction and first few chapters will likely be of the most interest to you and will help in understanding the leadership that is needed to bring a sustainable competitive advantage to your company.

The most difficult change is changing how you think and lead. Critical to your success is understanding what Demand Flow Technology is and why it will work for your company. The middle chapters contain the meat of the tools and techniques, with the specifics on how to accomplish the transformation. For the folks who enjoy getting into the actual mathematical models and calculations, the final chapters will be of greatest interest. This book will show a great deal of wear and tear as it is passed around and used. To help you be truly successful, it should not sit on the bookshelf in pristine condition. Remember that this change is a journey; and during a journey, your needs change. As you read and reread this book, you will discover something new each time you pick it up.

The Future

Recent news attention would suggest that outsourcing is a new phenomenon. However, even as early as 30 years ago, outsourcing was a common practice to leverage lower labor costs. The difference is that in today's market, outsourcing's destination has changed from Mexico to China. However, dramatic changes have occurred in the market to challenge a successful low-labor-cost outsourcing strategy. Labor is now less than 10 percent of product cost, and even the total elimination of labor would only marginally improve profits.

Companies struggle for a sustained competitive advantage. Global over-capacity and increased customer expectations due to the internet have shifted power to the customer like never before. No longer can a manufacturer forecast consumption, push inventory to market, or promote product to sell when the forecast is wrong. Now, a company must continually drive out latency from the entire supply chain — from the customer's customer to the supplier's supplier. Inventory may still count as a financial asset on the balance sheet, but most manufacturers have come to realize that inventory is a misuse of company assets and limits the ability to respond rapidly to changes in market demands.

Even process manufacturers, who by necessity must produce to stock due to the expectations of retail customer purchases, can benefit from the reduction of latency in the supply chain. To survive, all manufacturing companies need to become demand driven. This book can be considered the handbook in transforming from a push manufacturing environment to a demand-driven environment. Your future depends on it.

Introduction

The search for sustainable competitive advantage can seem like mission impossible. Scores of books by "thought leaders" champion breakthrough ideas—without the detail on how exactly to make it all work. Software companies sell a compelling vision, but every executive who has implemented any kind of software understands the effort that goes into making it all work. System integrators attempt to come to the rescue with best-practice implementations, but the risk of inadvertently sacrificing a competitive advantage is quite real.

The source of future profit is clearly different than it was in the past. With the advent of the internet, a company can no longer compete merely on its product's features and functions. Now it must provide a value to the customer, as well as drive profit internally. Adding to this challenge for the manufacturing company is the steady rise of manufacturing productivity and overall manufacturing capacity, without an accompanying rise in demand. Compared with the seller's market of just a few years ago, the world around manufacturing companies has turned upside down, and the customer has power like never before when it comes to pricing, product variety, and overall service. A transformation in thought and vision is necessary to survive. This transformation affects not only the manufacturing process but also employee, customer, and supplier relationships. In essence, a new approach has emerged for the entire supply chain.

History

Looking back just a short 50 years, computers would not be found in most manufacturing companies. The state-of-the-art manufacturer used order point as a way to control inventory. If a perpetual inventory were used, all receipts and issues were posted to manual cards. Manufacturers measured time in weeks and summarized time in months. The thinking was that if the company built extra product this week, that was fine — it would be used sometime in the near future. Labor comprised more than one third of product cost, and keeping labor busy and productive was the ultimate goal. This process was in use well into the 1980s.

Then, three visionaries — Joe Orlicky, George Plossl, and Ollie Wight — realized that if a future production plan could be made, material could be ordered to arrive just as it was needed, without stocking unneeded parts and without giving up labor efficiency.

The first commercially available material resource planning (MRP) systems came to market during this period, and a new industry was born. However, material was only part of the problem for manufacturing. As quickly as a computer with sufficient power to perform the calculations evolved, the calculations expanded to include the scheduling of work orders using move, queue, and setup times, in addition to actual work content. The elapsed time for these work orders was about 20 percent value-added time and 80 percent queue time. Still, companies mastering these techniques were able to outperform those that continued using order point. Inventories shrank, and visibility to provide commitments to customers improved. These tools continued to evolve with the integration of accounting, resulting in manufacturing resource planning (MRP II), and then further evolved to manage resources across an entire enterprise with the emergence of enterprise resource planning (ERP).

The most telling sign of the times was the development of advanced planning and scheduling (APS) systems that optimized scarce capacity and quickly rescheduled production when disruption occurred, which it regularly did. Over these 50 years, labor's portion of product cost steadily declined, eventually accounting for less than 10 percent of product cost today. Material cost now accounts for more than two-thirds of product cost, depending on the specific industry. Customers expect as a minimum a level of performance that was considered exceptional just a few years ago. The advent of the internet removed significant transactional friction.

Now, instead of just sourcing to known suppliers, companies can research and source globally with relative ease. A manufacturing company no longer has a future production plan that they trust to drive the material plan. In fact, inventory not used on an order may never be used in the future, because the customer may no longer be there. Even the practice of running promotions to push inventory to market may require the company to sell product at a loss just to eliminate the inventory. Significantly reducing the time between customer demand, production, and incoming supply provides benefits for each link of the supply chain.

The Solution

This different manufacturing focus requires a different manufacturing approach. The manufacturing process — from the product design to line design to supplier and customer communications — must enable the manufacturer to thrive on this new volatility. This is called Demand-Driven Manufacturing (DDM). Demand Flow® Technology (DFT) is a mathematically based approach that positions a company to become demand driven. DFT should not be confused with lean manufacturing or just-in-time (JIT) manufacturing. DFT is a trademarked technology for which its developer, John Costanza, won a Nobel Prize nomination.

Moving from the world of traditional "push" manufacturing to this DFT approach is no small change. The transition requires that everyone in the company change how they think. Increased incoming volatility now means increased competitiveness. No longer does a company need to trade off product variety, customer responsiveness, and profitability. Now a company can excel in all three dimensions. The traditional measurement system gets turned upside down in the process. No longer are efficiency, utilization, and absorption the three most important measurements. In fact, the more quickly a company embraces DFT, the faster and worse the traditional financial metrics become. Now a company focuses on throughput, investment, and operating expense. People (once considered a cost to be managed and reduced) become the most important part of the process, because they are the most flexible "machines" available.

This transformation is not an immediate wholesale change across the company. Becoming demand driven requires that the company adopt a continuous improvement, or "kaizen," approach. The kaizen events focus on those items constraining the company's ability to become demand driven.

The goal is not just to hold kaizen events; the goal is to focus them on where real benefit can result. This process may begin with engineering during the design process to ensure that products are designed to maximize flexibility or to eliminate error assembly. Remember that 80 percent of the cost of future production is committed during the engineering process. Any improvement in manufacturing (where 80 percent of the cost is incurred) can only marginally impact the total cost.

The process also might begin with sharing information with customers to understand more quickly what is needed and when. Very rarely does the process start with suppliers, unless the supplier is an integral part of the design process. Supplier collaboration usually follows manufacturing improvement.

Regardless of where the process begins, a synergistic relationship exists between the changes in business process and the technology applied. Both must change at the same time in a rapid, flexible manner for a company to be successful.

Just as production moves from big batches to small batches, improvement projects also become smaller. But when those improvements focus on overcoming the company's limitations, the bottom-line return is quick and significant. To become demand driven, the production batch becomes significantly smaller than the order batch.

Rather than having the first item in a batch wait until the last item is completed (as in traditional push manufacturing), DFT allows the manufacturer to produce within a continual flow of parts, allowing the batch to accumulate at the end of the process for delivery to the customer.

Performance Measures

As exciting as these changes sound, and as significant as their impact can be, a company's first step is to address performance metrics. Traditional measures of productivity (efficiency, utilization, and overhead absorption) encourage exactly the wrong behavior — to build inventory by keeping machines busy, even in the absence of demand. Economic production value is "earned" in a standard cost system when product is completed and goes to inventory. Some companies have switched to actual costing to attempt to capture more-accurate costs.

In reality, without bringing capacity and load into the costing equation, neither standard nor actual cost will provide accurate information. Throughput cost accounting provides a realistic picture of how an additional order will affect profitability and which products really are most profitable. If a company still feels the need to allocate overhead, it gains a much better picture by using throughput time, or time in the bottleneck resource, as the driver.

Measurements change for people, too. The entitlement mentality, where an employee is rewarded for the number of years on the job, is replaced with a program of paying people for knowledge. See chapter 10 for a discussion of "one up, one down" certification training and compensation. A critical concept in DFT is that the work force is something to be developed and rewarded, and not a target for persistent reduction and cutting. The goal becomes supporting a higher level of sales with the previous level of personnel. It is not uncommon for a company to produce 10 to 50 percent more with the same number of people once the company becomes demand driven. Remember that these people are all being paid today. If there is no need to add employees in the future, it follows that costs will not rise. To successfully execute this technique, it is imperative that sales not give away the advantage of operational improvement to customers. The point is to exploit the new capability to increase speed and provide a unique value proposition that enhances overall profitability.

Use of Technology

As technology evolves, so do the corresponding tools and techniques. From the earliest bill of material processor, to MRP, to MRP II, to ERP, and finally APS, the technology built on previous technology and evolved to the next level of capability. However, just as the Demand Flow transformation requires a different management thought process, the technology must also evolve from a different core. Even today, the production plan's gross-to-net calculations for optimizing inventory are the core of the most sophisticated ERP. In demand-driven companies, the core around which everything rotates is customer demand. This could mean building directly to order for discrete manufacturers or reducing the latency between a product that sells out of a retailer or distributor but returns to the manufacturer. A relative new technology, radio-frequency identification (RFID), can help reduce the latency in the supply chain by speeding more accurate transactions.

Remember the five rules for using technology:

1. Know the power of the technology. Determine what it will allow you to do tomorrow that is not possible today.

2. Define the value to the company of having this capability. Remember that technology can only provide value if it addresses a limitation that keeps the company from achieving its goal. Just as the kaizen events must address the limitation or constraint, technology decisions must focus there as well. Applying technology anywhere else in the company will only add cost.

3. Determine which business practices or rules help the company to work around that limitation. If the company is a going concern, then somehow it has figured out how to live without this capability. Unless the underlying business process is addressed, the improvement realized from the technology will be marginal at best.

4. Decide which business practice should be used now. Remember the synergy between business processes and technology. As technology changes, business practices must change. As business practices change, technology must change. No longer are implementation times measured in months. Now it is weeks, with return on investment also measured in weeks.

5. Identify ways to accomplish the transformation. This is all about change, and as we all know, change can be difficult. Once the company agrees on the problem, the direction of the solution, how the solution solves the problem, and how to manage unintended negative consequences, it is time to step up and lead the organization into the change. It is always surprising how quickly the change will happen when people understand what they are doing and why they are doing it, and share in the benefit to the company.

Summary

If this process sounds very overwhelming, that is good. You have begun to understand the magnitude of the process in front of you. However, do not be trapped into believing that you cannot make this change. Many companies have become demand-driven manufacturers and are not only surviving but also thriving in this new world of manufacturing. In this book, you will find a wealth of knowledge to aid you. Do not get caught up and confused by the acronyms. They are merely the tools required to build the future. First, articulate the vision of how your company can exploit a significantly faster response to the market to provide value for the customer at a profit. Next, look at the business processes in place today and the technology that supports them. Expect that both will change in a rapid and continual fashion. Only when a clearly articulated vision is aligned with business processes and technology can a company become demand driven and achieve the dramatic results discussed in this book. Remember that even the longest journey begins with a single step.

Dedication

*This book is dedicated to the spirit of freedom
and those brave men and women who risk their lives
preserving it each and every day.*

Acknowledgment

I was giving a presentation in Europe last year on Demand Flow®
Technology, when *The Quantum Leap* came up as a topic of discussion.
A gentleman in the audience stood up with the book in his hand and said,
"My goodness man, this is not a religion." I had to keep myself from
smiling. Knowing John Costanza, it is a religion — his religion.

John invented Demand Flow® Technology. It is the work of a lifetime,
and while the means of implementing Demand Flow have changed over
the years, only John can lay claim to being the "father of flow."
DFT is and will remain a significant contribution to the worlds of industry
and manufacturing.

John Costanza is a rare individual. His breed, unfortunately, is fading
away. We seldom find men of such great passion these days. He risked all
and gave all to his life's work of making companies the world over
more competitive. In the end, we have all gained from increased
competition, as it keeps us all on our toes. Manufacturing no longer
ranks as a secondary or tertiary organization within a company. In fact,
many companies now view it as a core competency.

The words "demand driven" embody the latest manufacturing panacea,
yet John showed that he understood the value of demand-driven
manufacturing better than anyone else in the years he spent
delivering DFT to the marketplace. A really good idea always finds its
way in the end.

John was single minded in his effort to bring DFT to the masses.
His early successes were quite meager, but he never gave up and
he never lost sight of DFT's future impact on our world. It is ironic that
someone would compare DFT to a religion, given that John resembled
a Baptist preacher traveling about and telling his story with great passion
to all who would listen, with his wife Linda supporting him all the way.

JCIT International would not be here, this revision of the book would not
exist, and many companies would not have realized the many benefits
of DFT without John's spirited passion.

I hope you will join me in wishing John the best in his retirement days
and join us all at JCIT in celebrating John as the true father of flow.

Dean Gilliam, Chairman
JCIT International

A Message from Dean Gilliam

"Go to where the People are..."

A Danish philosopher once said, "You must go to where the people are." In other words, to communicate effectively with your audience, you must relate to their needs. I hope you find this book meets that standard. We believe it will give the reader sustaining and substantive ideas.

Now and in the future, meeting customer demand on a real time basis is the first priority for manufacturing companies to survive, much less succeed. Large inventories of finished goods and the associated costs are plainly too prohibitive. Companies simply must have the manufacturing and supply chain processes in place to build to demand on a real time basis.

This book defines Demand Flow Technology (DFT) and explains the technology and the benefits associated with this award-winning business strategy of using manufacturing as a decisive competitive advantage. It is also crucial to acknowledge the many other types of tools and techniques in existence today so the reader may enjoy a more complete understanding of the world of manufacturing.

JCIT International's worldwide deployment of Demand Flow Technology has saved companies more than $7 billion to date and is responsible for saving manufacturing companies more than 4.5 million jobs in their home countries. Companies worldwide have used the tools and techniques of DFT to achieve success in their respective markets. We are honored to welcome you to the exciting universe of Demand Flow Technology.

The introduction to this book is by Carol Ptak, a world-renowned best-selling business author and our dear friend. Carol has always been on the cutting edge of manufacturing techniques and technology, and we are honored to have her involved in this book. I am sure you will find her comments highly enlightening.

Dean Gilliam, Chairman
JCIT International

About JCIT International

For over 20 years, JCIT International has trained over 90,000 manufacturing professionals from over 3,500 companies worldwide in the principles of Demand Flow® Technology (DFT). With state-of-the-art training facilities located in the United States, Europe and Asia, JCIT offers a variety of executive and management workshops.

Founded in 1984 by renowned author, industry leader and Nobel Prize nominee, John Costanza, JCIT International is committed to the development and delivery of the highest quality educational programs and implementation support programs within the manufacturing industry. Every workshop and consultation engagement is designed with one goal in mind: providing companies the highest level of customer service and a dynamic set of DFT tools.

DFT is a scalable, mathematically-based business strategy, specifically designed to allow manufacturers to respond faster and more efficiently to the needs of their customers and the marketplace. With a strong DFT foundation, companies control their supply, manufacturing and distribution pipelines based upon actual demand rather than inaccurate forecasts.

To learn more about Demand Flow® Technology, please contact:

US Headquarters
+1 303 792 8300
+1 800 457 4548
Fax: +1 303 792 8311
E-mail: info@jcit.com
6825 South Galena Street
Englewood, Colorado 80112 USA

European Headquarters
+45 7026 8300
Fax: + 45 7025 4402
infoeurope@jcit.com
Brendstrupgaardsvej 13
DK-8200 Aarhus N
Denmark

Asia Pacific Headquarters
+86 21 6335 2206
Fax: +86 21 6335 2209
infoasia@jcit.com
Level 41, Bund Center
222 Yan An East Road
Shanghai, China 200002

www.jcit.com

Chapter 1
A Global Perspective

The Changing Global Marketplace

The global marketplace has significantly changed the rules of business. Economic and political events have opened the doors to new competition. Free-trade agreements have leveled the barriers to new markets, and the global economy's growing thirst for new goods and services is continually accelerating the pace of competition. In this dynamic new environment, success no longer results from manufacturing the highest-quality products in the most cost-effective way. Companies must also respond to customer demand in real time. In other words, they must become demand driven.

Creative asset management, timely response to customer demand, and speed-to-market for new products are essential components of the demand-driven formula for success. Manufacturers that can't or won't adopt demand-driven techniques will quickly see their market opportunities fade away.

Speed-to-Market and Customer Responsiveness

As markets become more global and competitive, companies are striving to deliver innovative products to the customer as quickly as possible. They search for ways to engineer products that are ready for manufacturing, which can dramatically shorten the customer's wait time. One objective of this book is to demonstrate that by linking flow-manufacturing technology with simultaneous/concurrent engineering techniques, companies can reach new levels of responsiveness and speed-to-market. Manufacturing theory has always acknowledged the importance of speed-to-market. But what was once a high priority is now a competitive necessity.

Shorter product life cycles mean that the first company to deliver a product enjoys the highest profit potential. As competitors enter the market with similar offerings, the opportunity window and the profit potential quickly narrow.

More significantly, as product life cycles shrink, so does the time available to recover the manufacturer's initial investment. Products with long design, prototype, and preproduction processes may reach the market too late to generate sufficient revenue to cover development costs, let alone profit.

This accelerated pace penalizes companies that lack a rapid design-to-manufacturing process. These companies run a greater risk of incurring losses on each new product.

Repeated delays in product introduction ultimately jeopardize their competitiveness. In contrast, speed-to-market is a powerful tool that can elevate an average manufacturer to a position of market leadership and dominance. With speed-to-market, manufacturers are quicker to respond to the needs of their customers. They can better adapt to the rapidly changing forces of their market. They use as little working capital as possible. But most importantly, they create a sustainable competitive advantage.

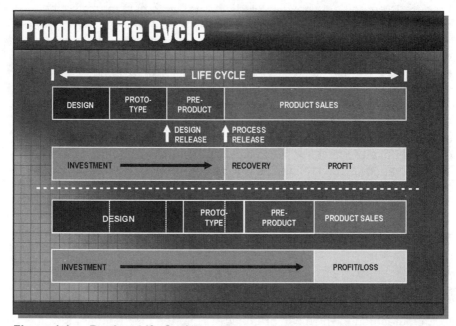

Figure 1.1 — Product Life Cycle

Two Different Approaches to Manufacturing and Design

Manufacturers typically approach manufacturing and design in one of two radically disparate ways. The first group relies on scheduled production based on work orders. They focus on functional product design and functional manufacturing. This approach and its variants are commonly referred to as materials requirements planning (MRP) and manufacturing resource planning (MRP II).

The second group embraces flow manufacturing, an approach that rethinks the importance of scheduled production. In 1984, John Costanza pioneered one such approach called "Demand Flow® Technology" (DFT). DFT is a mathematically based alternative to traditional, schedule-driven manufacturing. It is the most successful approach among flow and lean manufacturing models in combining mathematics and technology into a viable and successful business strategy. DFT manufacturing concentrates on process design and increased material throughput to bring product lead times in line with work content times.

Material and overhead elements constitute the bulk of product costs — as direct labor typically represents 3 percent to 10 percent of total cost. In most discrete manufacturing scenarios, material represents the majority of product costs, with overhead as the second largest portion. Direct labor accounts for the least. However, when evaluating ways to cut costs, manufacturers usually look to reduce labor expenses. As a result, many companies shift production to countries where labor costs are lower. These companies are, unfortunately, chasing the smallest portion of their product costs.

For years, manufacturers have focused on direct labor as the key to managing the manufacturing process. Material and components brought to the manufacturing facility are stored to await the available scheduled direct labor and machines. In many cases, companies bring in material earlier than they need it to ensure that production employees stay busy. This tactic increases on-hand raw materials and finished-goods inventories, which in turn increases the material and overhead portions of product cost. In fact, traditional costing and management systems can actually encourage this inventory buildup to create a misleading impression of labor efficiency and to absorb traditional overhead costs. Often, a product that takes weeks to produce actually has only hours of actual work content in it.

Tracking Process

Many Western companies have long employed complex scheduling systems to assist management in tracking products through a manufacturing facility as well as the efficiency of direct labor. MRP and MRP II were developed specifically to aid manufacturing planners and buyers in determining what to build and what to buy based on their manufacturing schedules. Companies maintained large inventories to support manufaturing processes that took days or weeks to deliver products.

In contrast, manufacturers in the Far East developed quality-focused flow processes that minimized material and overhead instead of direct labor. They developed techniques to replenish material as it was consumed using simple yet powerful pull systems.

Flow Process

Japanese manufacturers gained a competitive edge by emphasizing flow processes and their associated quality initiatives. Although many of these companies are now adopting the mixed-model techniques of DFT (originally patented by JCIT in the early 1990s), their original focus centered on minimizing lead time and maximizing productivity.

Shorter lead times and higher productivity are just part of the picture. The idea of building products to customer demand is the foundation of Demand Flow Technology. A true flow process allows companies to adjust volume and product mix on a daily basis. At the same time, DFT enables the highest possible quality levels, with minimum overhead and the lowest possible material costs. Companies pull materials into the manufacturing process and build products based on actual customer demand, not market forecasts.

Companies that implement Demand Flow® Business Strategy avoid the substantial inventory carrying costs borne by traditional manufacturers who maintain large inventories. Because the vast majority of product cost comes from material and overhead costs, Demand Flow manufacturers can produce products at substantially lower costs, even though their labor costs may rise. In many industries, Japanese labor costs are higher than in comparable American industries, but their overall costs are lower. Unfortunately, manufacturers in the United States and Europe face continuing pressure to reduce the direct labor and labor value in products. In response, many companies have moved their manufacturing operations to low-cost labor markets. Simultaneously, they carry large amounts of inventory in high-cost offshore facilities and invest large sums in transportation between the "low-cost" offshore manufacturing facility and the customer.

The Far East Dominates Employee Productivity Growth

Productivity growth per country is a telling indicator of which manufacturers will dominate global markets and where the competition will emerge. While total production output in the United States and Europe still leads the world, companies in the Far East have historically dominated real growth in output per employee.

Growth in output per employee in Japan is only a part of the success story. Japanese companies maintain high quality and low production costs while continuing to release innovative new products. Korean and Chinese manufacturing giants are adopting the same technology, tools, and techniques as their Japanese counterparts. And they're succeeding. While Japan has maintained growth in output per employee, Korea's growth exceeds that of Japan.

Other players are emerging from the Baltic countries and Poland, where labor costs are significantly lower than their European neighbors. The relationship of Baltic countries and Poland to the rest of Europe is analogous to the free-trade relationship between Mexico and the United States.

JCIT executives recently discussed the manufacturing business climate in China with a Chinese business leader who said, "The time to come to China for low labor rates has passed." Inflation is rearing its head in China, and regulatory constraints are finally catching up to early adopters of low-cost Chinese labor. If companies sell into China then it remains economically feasible to manufacture there. If not, the costs are far greater than they appear to be on the surface.

Nevertheless, corporate executives continue to announce their latest cost-cutting initiatives of outsourcing manufacturing to "low-cost labor markets" such as India or China.

In the end, these initiatives will likely result in profit lost to missed sales, because companies sacrifice the ability to respond to customer demand. We live in a demand-driven market, and there's no turning back.

A More Effective Strategy for Global Competition

The revolutionary tools of Demand Flow manufacturing provide a distinct competitive advantage in the emerging demand-driven market. Companies can manage their manufacturing pipeline based on actual sales rather than anticipated forecasts. Although originally developed as an executive level strategy, Demand Flow transforms the way entire organizations work internally as well as with their partners and customers.

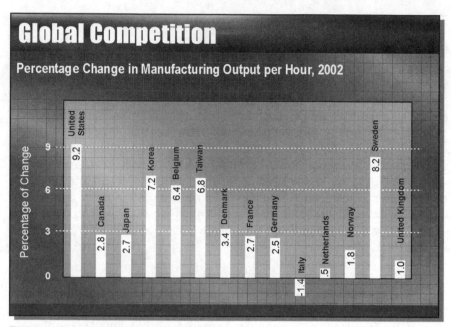

Figure 1.2 — Global Competition

Demand Flow is applicable to all manufacturing scenarios: discrete, process, continuous flow, repetitive, made-to-order, assembled-to-order, and any other model. Company size is irrelevant. JCIT's clients range from $2 million in annual sales to over $20 billion. Because the tools are mathematically based, they deliver results regardless of company size.

Having attained a manufacturing advantage with Demand Flow, a company's sales organization can offer customers shorter lead times. Companies can introduce new products more quickly. Engineering can simultaneously design new products using existing flow processes. Finance can take advantage of improved cash flow and lower inventory investments. Most importantly, the customer receives a high-quality product, delivered on demand. Ancillary benefits include a dramatic increase in inventory turns, a radical reduction of finished-goods and in-process inventory, increased productivity, reduced manufacturing floor space, and lower overhead costs. Simply put, costs go down and sustainable competitive advantage goes up.

To compete successfully in the global marketplace, manufacturing companies must consider the ability of each investment to deliver a richer product mix, greater flexibility, higher responsiveness to customer demand, and improved quality. With Demand Flow manufacturing technology, companies realize these advantages by linking production to actual customer demand. The following chapters detail the techniques, methods, and systems of the Demand Flow business strategy.

Chapter 2
A View to Manufacturing

A historical perspective on the evolution of demand-driven manufacturing begins with an understanding of material requirements planning (MRP). As mainstream manufacturing companies in North America adopted MRP, comparisons to Japanese manufacturing and its superiority to American and European models circulated widely. This productivity and quality gap has narrowed significantly over the last 15 years, due in part to the availability of better tools. However, the primary reason for the improvement in North America is the growing recognition that by meeting customer demand, manufacturing companies can realize increases in productivity and quality over and above what many lean manufacturing techniques strive to achieve.

MRP promised a wide range of benefits, including the ability to control and monitor manufacturing activities, track and record transactions, schedule and plan production, and procure parts. As more companies embraced the MRP methodology, they soon recognized its shortcomings. For example, MRP couldn't accommodate capacity restraints. Manufacturing resource planning (MRP II) then emerged to address the resource planning process, as well as other functional gaps in MRP. MRP and MRP II—combined with a variety of quality initiatives—sharpened the competitive edge for many companies. The next logical step in the pursuit of efficiency was to seek out low-cost labor. As a result, many corporations moved manufacturing operations to China, Poland, the Baltic States, and other burgeoning economies.

Even many Mexico-based manufacturers established a foothold in China in hopes of achieving lower direct labor costs. However, lower direct labor is not the path to market domination. The ability to meet customer demand in real time requires higher efficiency and productivity than simply lowering direct labor costs.

If candor reigned, most manufacturing experts would admit that the vaunted savings of "cheap" direct labor quickly evaporate after adding in the hidden costs for logistics, travel, inventory, and engineering.

In the initial writing of *The Quantum Leap*, John Costanza outlined these dynamics. He recognized that the underlying concepts of the Japanese flow methodology held significant potential for all manufacturers. The result was a reshaping of those techniques and methods into the solution we now know as Demand Flow® Technology (or DFT).

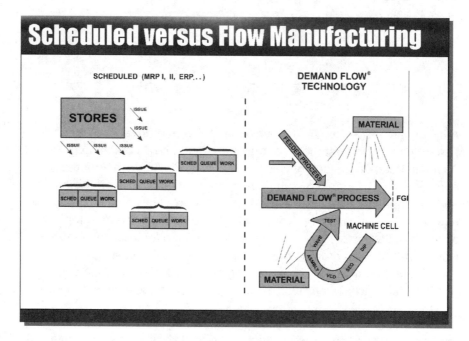

Figure 2.1 — Scheduled versus Flow Manufacturing

Core Methodology

Rather than endorsing a single, exclusive manufacturing strategy, most companies adopt a more pragmatic and eclectic approach. First, a manufacturing company should carefully decide on a fundamental methodology and then pick and choose among the available tools and techniques with an eye toward the ways in which they will support the overarching strategy. Many large companies already do this today and define this mosaic of techniques in their own terms and with their own processes.

Work-Order-Based Scheduling Systems

The work-order-based scheduling approach centers on a master production schedule (MPS) under the umbrella of an MRP/MRP II system. The methodology is comprised of a series of processes designed to arrive at a requirements plan based on the MPS. Work orders, routings, and their associated transactions are integral to this methodology.

Flow Manufacturing Systems

The first flow manufacturing system was the Toyota production system (TPS), founded by the Japanese auto industry during the 1950s. This technique was the first to use a line pulse, referred to as "takt (a German word meaning "beat" or "tempo"). Takt is calculated by dividing the time available to work by the number of customer orders expected each day. Work steps are then balanced according to the resulting takt in an effort to meet customer demand as quickly as possible with minimum in-process and finished-goods inventories. Methods such as "jidoka" (the Japanese term for the practice of stopping the production line when a defect occurs) help limit inventory exposure to high amounts of rework and/or scrap. TPS documentation indicates that if customer demand changes, significant challenges result.

As the contemporary response to TPS, mathematically based Demand Flow Technology creates a flexible flow of products, thereby removing manufacturing as a variable in meeting customer demand on a daily basis. Manufacturing operations accommodate mixed-model products down the same manufacturing line, in varying quantities, based on changes in customer demand. The material resupply processes—both internal and external (i.e., the supply chain)—link mathematically to the mixed-model production environment. The facility responds to actual customer demand from a flexible environment that adjusts mix and volume each day through mixed-model lines and cells. Support organizations incorporate these same tools as appropriate, making the production area more predictable and easier to understand and manage. DFT also provides support organizations the capabilities to contribute to the primary DFT objectives of speed in response, increased parts-per-million quality, and lower cost—ultimately freeing up working capital for future investment, growth initiatives, and higher profitability. All these benefits help explain why DFT has been called the "ultimate form of lean manufacturing" and "the science of pull."

Tools and Techniques

Kaizen

The kaikaku and kaizen methods (roughly translated as "radical improve-ment" and "continuous incremental improvement," respectively) focus on the manufacturing environment, though their benefits aren't limited to the factory floor. Typically, these methods address a limited scope of manu-facturing activity, even down to individual operations in the factory. A ded-icated team concentrates on optimizing all operations and processes of a given area. However, these process improvements don't address the impact of the targeted operation on the other areas of the facility. In fact, the resulting optimization in one area can negatively (though unintention-ally) affect response, quality, and cost in other areas of the supply chain.

Lean

Lean is a manufacturing philosophy first introduced in the TPS and later elaborated on by James Womack in his book *Lean Thinking*. Today, lean manufacturing is synonymous with the application of myriad productivity, material movement, and housekeeping systems that encompass produc-tion, quality, maintenance, and engineering. While lean principles and methods originated with TPS-based flow methods as applied to the auto-mobile industry, a wide range of concepts—including value-stream map-ping—now fall under the umbrella of lean manufacturing.

Value Stream Mapping

A value stream is the sum total of actions required to bring a product through the manufacturing process (or a subsection of that process) from start to finish. The goal is to identify and eliminate the waste in the proc-ess, with "waste" defined as any activity that does not add value to the final product. A "current-state" map defines initial conditions, with a "future-state" map outlining the ultimate goals. The process of mapping out the activities in the production process—including cycle times, down times, in-process inventory, material moves, and information flow—helps companies visualize the current state of their processes and guides them toward the future desired state. Value stream mapping provides only the view, not the tools needed to achieve the future-state results.

5S

Based on five Japanese words that begin with "s," the 5S philosophy focuses on creating an effective work place. The intent of 5S is to simplify the work environment while improving efficiency and safety through ongoing "housekeeping." The 5Ss are:

Sort (Seiri)—eliminating unnecessary items from the workplace.

Set In Order (Seiton)—creating efficient and effective storage methods.

Shine (Seiso)—maintaining a thoroughly clean work area.

Standardize (Seiketsu)—standardizing best practice in the work area.

Sustain (Shitsuke)—defining a new status quo and standard of workplace organization.

Theory of Constraints

The goal of the theory of constraints (TOC) is to improve efficiency by identifying and eliminating bottlenecks with tools such as drum-buffer-rope (DBR) scheduling. Constraints are the critical areas that limit the throughput efficiency of the manufacturing system. DBR operates on the theory that a single constrained resource usually limits overall production rate. The DBR approach manages the constraint first, and orchestrates all other activities and schedules according to that restraint.

Six Sigma

The goal of Six Sigma is to increase profits by eliminating variability, defects, and waste. Six Sigma is a rigorous and systematic methodology that uses statistical analysis to measure and improve a company's operational performance, practices, and systems. It identifies and prevents defects in manufacturing and services to anticipate and exceed stakeholder expectations.

TQM

Total Quality Management is a structured system for satisfying internal and external customers and suppliers. It accomplishes this by changing an organization's culture through an emphasis on continuous improvement in all facets of the business.

Supporting and Sustaining Software Tools

Enterprise Resource Planning (ERP)
An ERP solution is a business management information system designed to optimize effort and resources across the various departments and locations within a business. It provides an integrated source of data and automated processes to effectively plan and control all resources needed to take orders from customers and manufacture, ship, and account for those orders.

Supply Chain Management (SCM)
A supply chain is a network of facilities and distribution providers that performs the functions of procurement, the manufacture of intermediate and finished products, and the distribution of finished products to customers. Supply chains exist in both service and manufacturing markets, although the complexity of the supply chain may vary greatly from industry to industry. SCM comprises the design, planning, execution, control, and monitoring of supply chain activities, so that supply is synchronized with demand to create value for all participants.

Advanced Planning and Scheduling (APS)
APS systems employ constraint-based scheduling capabilities to optimize supply chain resources and capacity throughout a long-term planning horizon. They analyze the material requirements and plant capacity (within the business and across the company's partner network) to define a manufacturing plan intended to make the best possible use of the supply chain resources, with detailed production schedules that balance customer requirements with the manufacturer's fiscal and operational production targets.

Manufacturing Execution Systems (MES)
An MES helps manufacturers operate as efficiently as possible by tracking the flow of information on production schedules, inventory availability, and work in progress to and from the shop floor. When implemented correctly, an MES can improve the management and reporting of each of the primary production activities in a plant.

Product Data Management (PDM)

PDM systems are designed to allow companies to better manage product information, from initial design through engineering and manufacturing. Companies realize increased control over the change management process and can better track product evolution and ensure the manufacturing feasibility and ultimate profitability of product design.

What Is World-Class Manufacturing?

World-class manufacturing is a philosophy that unites a range of techniques to achieve significant continuous improvement in manufacturing performance. It is a process-driven approach that addresses quality, cost, delivery speed, reliability, flexibility, and customer service as functions of success.

The Case for DFT

A thorough understanding of Demand Flow tools and techniques quickly highlights the true power of this technology. DFT encompasses all of the effective attributes of other flow systems and methodologies, and combines the best of the various approaches and techniques used in manufacturing today. Most significantly, DFT accomplishes these objectives in a clear, pragmatic, and systematic fashion that transforms an entire business. It provides a common language across facilities and departments that allows companies to become demand-driven manufacturing success stories. Once a company has achieved manufacturing success with DFT, it often uses the same tools and methods to reduce lead times and improve efficiency in other areas of the business, such as engineering, finance, and administration.

Chapter 3
Strategy and Technology
for World-Class Success

Scheduled Versus Flow Manufacturing

The differences between scheduled manufacturing and Demand Flow manufacturing are substantial, significant, and many. Among them are fundamental differences in strategy, objectives, methodology, techniques, and the utilization of people. They also differ in their underlying goals. Work-order-based, scheduled manufacturing strives for high productivity and strong performance on key tracking metrics. In contrast, Demand Flow manufacturing aims for demand-driven production with minimal in-process inventory and the highest standard of process capability. Essentially, a flow manufacturer can produce in hours or days what may take weeks for a scheduled manufacturer to produce. And the flow manufacturer does this with greater manufacturing output, substantially higher quality, reduced work-in-process dollars, less workspace, reduced scrap and rework materials, increased labor efficiency, and reduced material costs.

Scheduled Production Philosophy

Work-order-driven manufacturing plants are typically designed around functional production departments. Also, they usually have a large store-room for raw materials and subassemblies. Production follows the scheduling of a fabricated part or subassembly.

These items are then routed from functional department to department based on the product's or subassembly's scheduled batch or lot quantity. Functional work centers and departments arrange their machinery and assembly areas to meet the requirements of this routing.

For example, several similar punch presses will be grouped together into a single functional press department work center. This functional arrangement could also apply to functional subassembly and test areas (see Figure 3.1).

In scheduled manufacturing, raw material waits in the storeroom. Once the assembly or fabricated part is scheduled, a work order is released, and the material required to produce the assembly is issued based on a planned production start date and start quantity. All required material is issued according to the quantity of fabricated parts or subassemblies scheduled for production. After the production item is scheduled and material is issued from the storeroom, the kit, or grouping of material is placed into a queue at the workstation or work center where the production will take place. The scheduling and issue of material usually takes between a few days and a week. When the scheduled production item becomes top priority at the work center that will produce the item, the kit of material is removed from the work center queue and placed into production. The kit, or group of parts, moves from operation to operation until the fabricated part or subassembly is completed. Work orders track material usage and labor efficiency, and variances are reported when the work order closes.

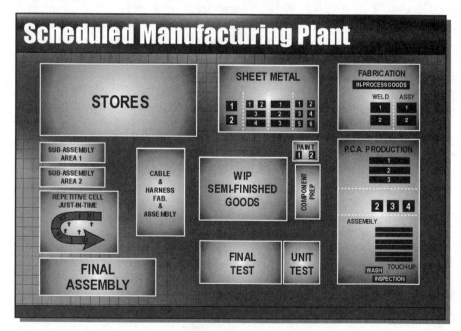

Figure 3.1 — Scheduled Manufacturing Plant

Quality Exposure of the Lot

Once the scheduled quantity or lot of parts is complete, it usually goes to an inspection area for quality verification. If the parts fail inspection, they return to the production department for rework. Since all parts were produced in a batch at the same time, the quality defect exposure level affects the entire lot. In other words, it is very common for an entire lot to share the same defect. Lots can then be processed through the storeroom using another work order that routes the fabricated part or subassembly onto another work center or workstation that performs the next higher level of subassembly or fabrication. This process continues until the manufactured item or subassembly becomes the top-level product, at which time it is ready to ship to the customer. This multilevel or subassembly manufacturing approach typically takes weeks to produce a product that may require only a few hours of actual machine and labor content.

Lead Time Inflates Inventory

The number of days required to complete a product—from the initial scheduling of the first, lowest level manufactured item or subassembly to completion of the final product—is defined as manufacturing lead time. Each machine and workstation in the production process adds its own process queues and waiting time. Subassembly items and kits follow the same production flow; they are produced by functional work centers, then stored, then issued according to the final product's production schedule.

Traditional scheduling techniques rely on unnecessary queue and raw material storage in anticipation of production. Because of the extended production process, companies create large inventories of finished goods to satisfy fluctuating customer demand.

The manufacturing scheduling process utilizes subassembly-manufacturing techniques, along with a multilevel bill of material. Because of rigid scheduling and its required lead times, this traditional manufacturing technique requires a substantial amount of inventory in process at all times. Additionally, manufacturers who employ these scheduling techniques realize 10 inventory turns or less per year, on average. The resulting increase of manufacturing lead times can complicate forecasting, given that the longer the manufacturing lead time, the more difficult it will be to predict customers' future needs.

"Scheduled" Manufacturing and Customer Responsiveness

Put simply, scheduled manufacturing lead times make it difficult to react quickly to changing customer demand. The combination of long manufacturing lead times, queues at each workstation, and frequent trips to the storeroom increase the time between the customer's order and manufacturing's completion of that order. While the manufacturing process is fixed and inflexible, the customer's position may not be. Indeed, customer demand often changes before the manufacturer completes production. If the customer's expectation for order lead time is shorter than the manufacturer's actual lead time, the manufacturer must carry larger, capital consuming finished goods inventories.

The longer the manufacturing lead time, the more likely that customer orders and sales forecasts will change in the interim. Such changes are frequent in traditional manufacturing, and they require major ongoing changes to the actual production plan. While manufacturing personnel often complain about changing sales forecasts, the problem originates with the length of time required by the manufacturing process. The longer the actual manufacturing takes, the greater the likelihood that sales will require a change to scheduled production to address changing customer needs.

Another distinct characteristic of schedulized manufacturing is that without a schedule and issued parts nothing happens in production. Also, while production employees are vital to specific workstations an machines, they have little responsibility or authority in the overall production process.

Extensive Storage Systems

Parts handling and storage represent two focal points of scheduled manufacturing. Raw materials are received, inspected, and stored. Then, based on a schedule, material moves to the production process. Often, materials return to storage in the form of subassemblies or fabricated parts before becoming part of higher-level subassemblies or the finished product. These nonproductive receiving and storage systems require significant control and monitoring during the production process, contributing to unnecessary overhead costs.

Flow-Manufacturing Foundation

DFT is a much simpler but incredibly powerful manufacturing technology. It focuses on a more aggressive flow process that seeks to minimize or completely eliminate any non-value-added work in the production process. It also emphasizes quality at the machine and production employee levels. DFT's primary objective is to build a high-quality product in the shortest production time at the lowest possible cost. The flow process is comprised of a sequence of tasks, and views the finished product as a "pile of parts" instead of a complex multilevel bill of material.

The production lines in a typical flow-manufacturing facility may have multiple feeder lines and/or machine cells feeding into it, with no distinguishable sub-assemblies in the production flow. The machine cell is a group of dissimilar machines needed to produce a family of similar products. The process does not consolidate milling machines, for instance, in one location, drill presses in another, and trimmers in yet another. In flow manufacturing, machine cells consisting of different types of equipment are arranged with just enough of each to maintain a steady and efficient flow to feed the consuming production line.

Products Flow On Demand

Where fabrication of an item or assembly is required prior to entering the main line, feeder lines attach directly to the main line where needed. As such, flow manufacturing requires minimal external subassembly work and no storage for completed work. Feeder lines are flow processes at the point where they are consumed.

Total Quality Control (TQC) methodology regulates the production process and serves to minimize production defects and maintain quality throughout the process. Easily understood, pictorial TQC method-sheet documentation guides employees through the work content and TQC verification processes.

Flow manufacturing entails a high quality production process based on a daily production rate, with the rate and mix varying day by day. Daily fluctuations occur in response to actual customer demand. The combined fluctuation of rate and mix can be quite significant, although spread over several days. Flow manufacturing eliminates queuing, waiting, and in-process scheduling. It also utilizes a demand-pull process, where product is pulled from the back of the process in a flexible and reactive fashion—reaching from product shipping back through the entire production process and out to external manufacturing suppliers.

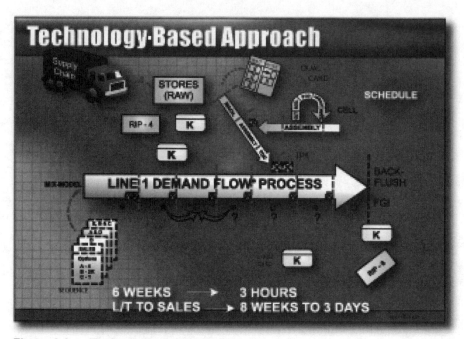

Figure 3.2 — Technology-Based Approach

Flexibility and Demand Pull

Demand Flow manufacturing techniques work with any product, high or low technology, and machine- or labor-intensive environments—once a company has overcome its resistance to changing outdated practices. The production process continues to function well at lower volumes, even when up to half of the production team is absent. Demand Flow manufacturing technology is a total business strategy that crosses all organizational boundaries and requires a commitment to revolutionary change from all top managers.

The Demand Flow manufacturing methodology adapts to change, adjusts to the planned or unexpected absence of multiple employees, and responds to market fluctuations because it fosters short reaction times. It is far more capable than traditional manufacturing of accommodating major changes in demand in a shorter period of time.

Flow manufacturing utilizes a demand-pull technique to communicate the demand to build product as well as the demand to pull material to replace that which has been consumed in manufacturing products. The demand-pull material system utilizes a technique called "kanban." Kanban is a Japanese word meaning "communication signal" or "card."

Kanban communicates a demand to produce a manufactured item and to pull materials into the manufacturing process without scheduling. More than just a communication card, kanban is an essential component of flow manufacturing: kanban encompasses multiple techniques, such as single card, dual card, and multiple card kanban systems. Kanban—and the cross-trained people who work in flow manufacturing—are key elements in maintaining the flexibility and responsiveness of the TQC process.

Process Pull Before Supplier Pull

Initially, converting to flow manufacturing does not eliminate the central storeroom. However, that is the ultimate objective. Although the storeroom is an essential part of the traditional scheduled-manufacturing philosophy, it is not a requirement of demand-pull flow manufacturing. Besides eliminating the storeroom, another major objective of flow manufacturing is to improve total inventory turns to a minimum of 20 per year for competitive survival—and 30 to 60 turns per year soon after.

Initially, parts will continue to come from suppliers with track records of questionable quality and late deliveries. As the part in question becomes a priority, a flow manufacturer works with suppliers to produce higher-quality parts as verified by the supplier's process at the supplier's location. Once the quality of a part has been ensured, the manufacturer's next priority is to establish consistent, on-time deliveries. By creating freight networks around proven suppliers, manufacturers can secure the delivery of quality materials on a more frequent basis without any increase in freight costs. However, suppliers continue to bear the burden of verifying quality before materials leave their docks. These quality initiatives will stream directly into the flow-manufacturing production process, alleviating the need for additional inspection of incoming or stored items. An area called "raw in-process" (RIP) inventory can be established to position parts in close proximity to where they will be used. On an average, RIP contains approximately one week's worth of raw parts. Over time, the number of parts delivered directly to RIP increases, and the number of parts delivered to the central storeroom decreases.

Focus on Quality, Costs, Turns

The objective of Demand Flow manufacturing is not to dismantle the traditional storeroom and scatter its material throughout the production process. The objective is to focus on quality, linear manufacturing, demand pull, and on-time delivery to allow quality parts to flow directly into the production process. Central storerooms are not essential to a Demand Flow manufacturing environment.

Without the scheduled products and subassemblies of traditional manufacturing, the raw-material storeroom is unnecessary. Flow manufacturing seeks the highest possible number of inventory turns and lowest possible overhead. The Demand Flow manufacturer works closely with, but not at the expense of, suppliers to achieve quality parts, on-time delivery, and lower costs as a result of DFT communication technology.

Total Customer Satisfaction

Many converts to Demand Flow manufacturing are originally drawn to the technology because it is market driven. It is guided by and responsive to changing sales requirements at a level beyond the traditional scheduled philosophy.

Since the Demand Flow strategy drastically reduces the manufacturing lead-time, sales personnel can make changes in a shorter period of time without negatively affecting the production process. However, management must be committed to frequent and continuing input from sales. Although the daily manufacturing rate can adjust to meet varying customer demand, changes must fall within a predetermined flexibility agreement negotiated among sales, manufacturing, and material suppliers.

Broadening Horizons

Demand Flow manufacturing facilitates a more successful selling effort as a result of reduced lead times. The sales department assists in the production process by contributing to market forecasts and finished-goods management. With flow manufacturing, sales can pursue opportunities that were previously beyond the company's reach.

Flow manufacturing is a company-wide, total business strategy that begins with the key members of the management team. In addition to a closer relationship with sales, greater flexibility, and supplier absorption of quality-control functions, flow manufacturing also encourages a closer working relationship between production and the finance, research and development, design engineering, and human resource teams. Throughout the company, roles change and departments expand beyond their normal functional niche. With Demand Flow Technology, manufacturing becomes a competitive advantage in competing globally.

World-Class Demand Flow Manufacturing

To become a world-class Demand Flow manufacturer capable of competing in the global marketplace, companies need to master these interdependent and crucial elements:

1. *A company-wide Demand Flow business strategy that can be fully supported by management.*
2. *A demand-driven manufacturing flow process sequenced according to customer order activity without production scheduling.*
3. *Production volume and mix adjusted each day, based on actual customer demand.*
4. *Financial management that is consistent with the flow process and labor tracking.*
5. *Simultaneous engineering to design products and processes concurrently.*
6. *Total employee involvement focused on the technical perfection of the product and process sequence of events.*
7. *A commitment to success by the entire organization.*

The business strategy includes a total commitment from management to meet customer demand using flexible, total quality flow technology. Customer satisfaction must reign supreme, as the corporate goal is to produce the highest-quality products with the quickest response to customer and market demands.

The company-wide DFT initiative includes TQC, engineering design, linear daily rates, process quality, operational cycle time (or takt time), flexible material forecasting, flow process costing, and backflush management analysis. A flexible staff must focus on quality and abandon traditional productivity goals. Production employees enforce quality through the use of operational method sheets. They also assume responsibility for defined assembly, fabrication, verification, and total quality validation. They must learn at least two other operations in addition to their primary responsibility, and can swiftly fill those positions to meet adjustments in production volume.

Flow manufacturing views people very differently than traditional manufacturing. Traditional scheduled manufacturing emphasizes cutting labor costs, even though labor costs for most efficient manufacturers have dropped to between three and ten percent of product costs. Flow manufacturing focuses on the other 90 to 97 percent of product costs—materials and overhead—while emphasizing employee involvement in building a high-quality product using a continually improving process.

Evolution or Revolution in Manufacturing

The benefits of Demand Flow business strategy can provide a corporation with a quantum leap in the competitive ability to produce high-quality, low-cost, and customer-responsive products. Flow-manufacturing techniques are fundamentally opposed to many of the techniques of schedulized, departmental, and subassembly production of traditional MRP II manufacturing. Progressive corporations hoping to achieve the unique distinction of world-class manufacturer must revolutionize their company-wide business strategy and technology.

Manufacturers who attempt to evolve toward the benefits of flow manufacturing (MRP II—lot size of one) without adopting the fundamental flow-manufacturing techniques, methods, and systems risk significant challenges that ultimately minimize the competitive benefits. The transition to flow-manufacturing technology is a challenge, but the benefits are enormous for those with a leadership commitment and desire to be the best.

Chapter 4
Flow Technology for
World-Class Manufacturing

The world-class manufacturer focuses on creating a high-quality and extremely efficient flow process that responds to customer demand. In the past, the traditional manufacturer didn't focus on the production process, despite its importance in determining the quality of the final product. With this in mind, manufacturers should recognize the strategic importance of the production process.

As a company evolves toward world-class flow manufacturing, the nature of the production process, along with key performance measures, will change significantly. Traditional terms such as scheduling, queue, labor tracking, subassemblies, production routing, and others become less relevant, if not altogether unnecessary.

Flow Manufacturing's New Terminology

A new vocabulary has emerged to describe flow manufacturing, and many companies are now using these new terms. The flow manufacturer employs terms such as TQC, sequence of events, team pass, kanban, total product cycle time, RIP, linearity to plan, flexible forecasts, cycle-time costing, operational cycle time, takt time (the German word for "rhythm" or "beat"), and many others.

DFT builds on a production flow process that uses a range of kanban tools to pull material into and through the production process as material is consumed. The rate-based production flow pulls material from a nearby point of supply. This flexible pull environment views a product as a "pile of parts" that flows through a sequence of events, ultimately resulting in the final product. The underlying objective of DFT is to produce the highest-quality product in a customer-responsive flow process.

Building a Demand Flow Process

Demand Flow manufacturing is a pull process in which demand pulls materials from the back, or the completion, of the product. Pull begins at the very end of the production-flow process and continues upstream through the manufacturing flow, through feeder processes and machine cells, through usage inventories, and eventually back to component and material suppliers. Customer demand pulls material and parts through the process, with the daily rate achieved at the end of the process.

This methodology differs from the scheduling and lead-time techniques of traditional manufacturing. Product synchronization (see Figure 4.1) defines the relationships of the individual flow processes coming together to create the part or product. Thus, the flow process may resemble an inverted tree composed of individual sub-processes, with assembly or machine cell branches feeding into the main flow at the points where the components manufactured in these feeder processes are needed. Once the product synchronization is defined for a product, each of the individual processes defined is broken into a TQC sequence of events (SOE). Because the end of the process has the highest priority in a DFT implementation, defining the TQC sequence of events begins with the final assembly processes.

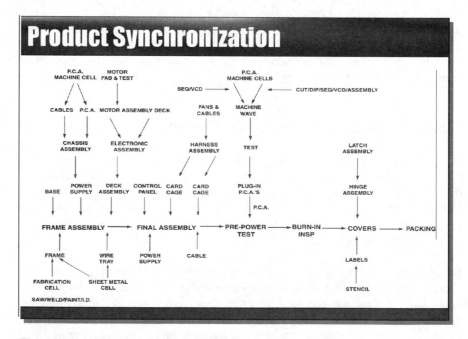

Figure 4.1 — Product Synchronization

Sequence of Events

The SOE is a DFT tool that describes work tasks and associated quality checks (See Figure 4.2). In this instance, TQC stands for total quality check and identifies the correct method when multiple methods of performing a work step exist. An SOE specifies the series of work content steps and the TQC criteria to ensure quality product. When performing a work sequence, traditional manufacturers tend to think in terms of batches, lumps, and conventional subassemblies moving in and out of storage-holding points as part of the manufacturing process.

This thinking should be recognized and avoided. Instead, manufacturers should think in terms of the natural flow of the product through the processes defined in the product synchronization. In some industries this involves following of a product batch rather than a single product. Nevertheless, the required thinking is identical.

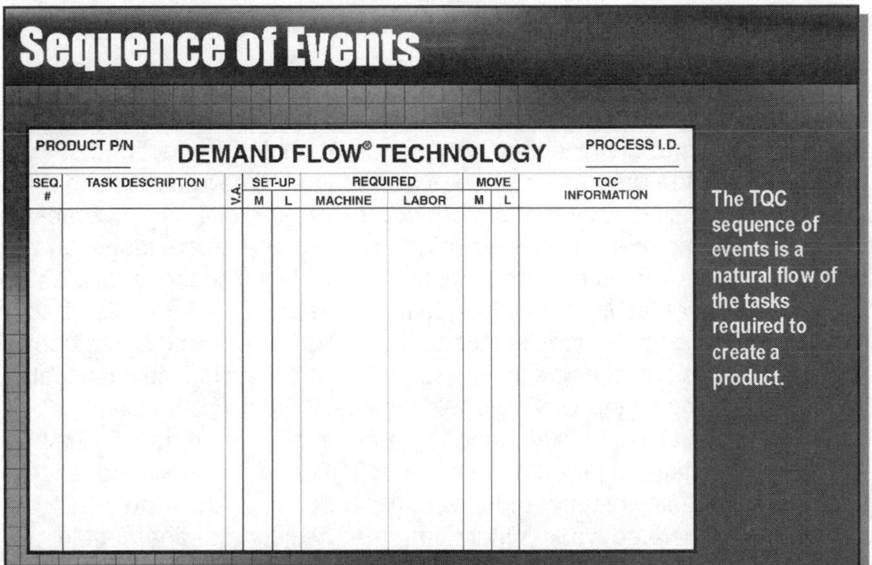

Figure 4.2 — Sequence of Events

Categories of Work

The sequence of events is a natural flow of tasks required to create a product. It describes the sequence of work and the quality criteria for each work step. Each task in the sequence of events falls into one of four categories of work.

The four categories are:

1. Required labor work

2. Required machine work

3. Setup time

4. Move time

The quality requirements for each step are then identified. The manufaturer cannot achieve the primary objective of producing the highest-quality product without understanding both the specific work steps and the corresponding quality requirements. Above all, the manufacturer must commit to enforcing the highest-quality standards among the people and machines that build the product. The path to total quality begins with a total-quality process.

In-Process Quality

World-class manufacturing starts with a flow process whereby the people who build the product have the responsibility, the TQC tools, the authority, and the methods to achieve their goal. External quality-inspection techniques—where the product undergoes inspection at specific points in the process, often including at the end of production—are still current practice in many companies. These techniques focus on external inspection tools and final product testing to identify failures. Many of these failures can easily be avoided by enforcing quality at the process level. Manufacturers must overcome the resistance to a new TQC process that replaces a "broken" process responsible for inferior parts and products. The comments "We've always done it this way" and "We're unique" typify this resistance. Quality must start in design engineering and remain the primary focus at each step of building a product. With the definition of the TQC sequence of events and quality criteria, the flow of the product dictates the line layout. The associated work content time also assists management in determining the number of machines and people required to produce the projected volume for a given production line.

Classification of Work

Every step to manufacture a product falls into one of the four categories of work. The process of classifying work ensures that the product specifications are understood and met, and it prioritizes improvement to the process. While all manufacturing steps fall into one of these four categories of work, not all work adds value to the product or process, even though it is critical to meeting the customer's expectations and product specifications.

Value-Added and Non-Value-Added Steps

Each step is either a value-added or non-value-added step. Value-added steps increase the worth of a product or service for the customer. Value-added steps can only be determined from the customer's perspective. It is essential to characterize each step as either "value-added" or "non-value-added" and to increase the number of value-added steps whenever possible.

Sometimes it isn't possible. For example, in-process testing not required in product specification is not value-added work, even though testing makes good business sense. Testing is preferable, given the process variables and material inconsistencies, but it does not add value. Testing falls under the "setup time" category. Testing would be value-added only if the customer required it as part of the product's specification. The product needs to meet performance and reliability standards, and in-process testing is a way to prevent process and material defects from reaching the customer. But the testing process itself does not add value.

Essential Labor and Machine Time

Required labor time comprises the steps performed to meet published product specifications. Although meeting product specifications requires labor, not all labor time is value-added.

Likewise, required machine time comprises the essential steps performed by machines to meet product specifications. As with required labor time, required machine time may or may not add value to the product.

Move and Setup Time

"Move time" is the time spent in moving products or materials from the point of production to the point of consumption. Move time may include labor and machine time. It is always considered non-value-added work. Appreciable move time usually indicates a poor production-line layout.

"Setup time" is work performed prior to required machine and labor time. It, too, is always non-value-added. Setup time can range from time spent changing a tool pack and making necessary adjustments on a large machine to time spent opening and removing a cable from a package. Once the non-value-added step is identified, modifications in packaging, line layout and machine-setup procedures can often reduce setup time.

The use of a TQC sequence of events to document and understand long, complex setup requirements provides considerable improvement in reducing required steps, improving consistency, eliminating steps, and simplifying setups.

Departure from Traditional Routing

The TQC sequence of events is quite different from traditional product routing. Traditional product routing tends to be of a summary nature and typically includes operations for assembly, inspection, testing, setup time, move time, and run time for machine and labor. The traditional routing is useful in moving product from work center to work center and in scheduling the planned hours in each department or work center. Labor routing does not distinguish between a value-added and non-value-added step. Thus, conventional manufacturing offers no effective way to determine which steps to eliminate. The traditional router gathers employee efficiency data and process-performance data based on the scheduled work. Most important, the traditional router does not contain the specific verification and TQC criteria that are essential to a total-quality process. Typically, a traditional router directs the product or subassembly to quality control for approval by an external inspector.

The TQC sequence of events is the key element in the design of a flow process. It formalizes the process. It assists in calculating total product-cycle time. And it creates a framework for process improvement by identifying defective steps and eliminating non-value-added steps. Standard routings have little value in flow manufacturing and should not be used in the transition to Demand Flow Technology. Any compromises during the initial definition of a quality flow process subsequently affect the success of the overall project. Such compromises often result from management's lack of understanding or commitment.

Total Time Calculations

Once the TQC SOE has been defined, the total time to build the product can be calculated. The sum of all machine, labor, setup, and move sequences is the total time to build the product. The total time to build the product is usually analyzed as total labor time and total machine time. Labor and machine time help determine staffing and machine utilization based on the expected daily rate volume. All work content (value- and non-value-added) combined reveals the total labor and total machine time to create the product.

Every value-added step is required to meet customer expectations and manufacturer specifications. All value-added steps fall within a cost category that charges them directly to product costs. The non-value-added steps fall into a cost category for ineffective-manufacturing costs. These "ineffective" costs don't result from customer requirements or product specifications; they are inherent in the manufacturing process. Thus non-value-added steps contribute to higher product costs and lower profit margins. The relationship of value-added time versus total time yields the following process-efficiency formula:

Process-Efficiency Formula

Process Design Efficiency % = VW x 100 / TT

TT = Total Labor Time + Total Machine Time
   (VW + NV)

VW = Sum of the Value-Added Work Content (Machine and Labor) Time

NV = Sum of the Non Value-Added Work Content (Machine and Labor) Time

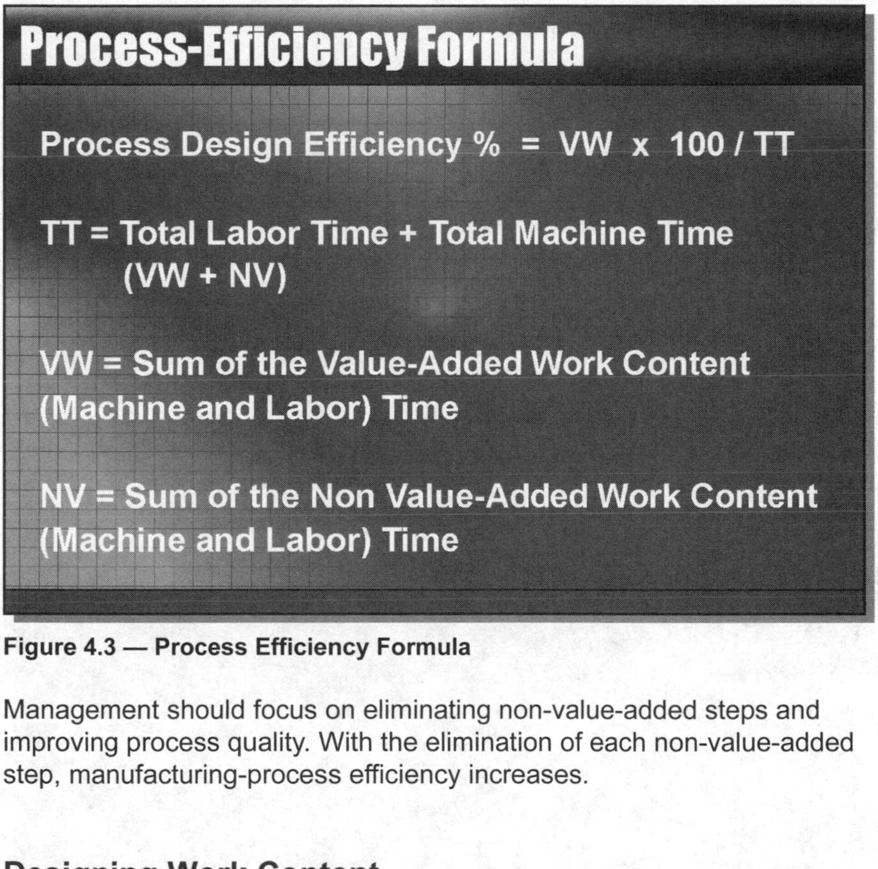

Figure 4.3 — Process Efficiency Formula

Management should focus on eliminating non-value-added steps and improving process quality. With the elimination of each non-value-added step, manufacturing-process efficiency increases.

Designing Work Content

The TQC SOE identifies the total work and total quality criteria to build a product. Once identification is complete, the manufacturers can begin designing a flow process by grouping work content into equal pieces of work. Under ideal conditions, each piece would require exactly the same length of work content time. An ideal layout of the entire production process—including the line, feeders, and machine cells—would show each process cut into equal pieces of work content time. If creating a product took a total of 16 hours and production involved 32 increments of 30 minutes each, the pull process of flow manufacturing would function smoothly, perhaps perfectly, completing a product every 30 minutes.

However, since a process relies on imperfect people and machines, absolute synchronization is impossible. Therefore, companies employ a series of balancing techniques developed for flow manufacturing. These techniques have the effect of "equalizing" pieces of work. They shape the relationship of processes and coordinate the elements of work content. Operational cycle time (the takt of the process) and flow-balancing techniques (which drive the design of the entire process) are two such techniques.

Line Design for Flow Manufacturing

Demand at capacity (daily) and the corresponding flow targets must be defined for each manufactured product. The targeted daily rate (demand at capacity) derives from sales and management agreement. Normally, flow lines are designed one time to accommodate the highest required demand at capacity (daily). Usually, that is a volume that cannot be surpassed without employing a second or third shift or extending the workweek to six or seven days. Although they are designed for a specific volume (capacity), flow-manufacturing lines are flexible and can easily run well below that volume. As actual demand changes, the range of volumes produced will fall between the designed maximum volume and 50 percent of that volume, without redesigning the line or changing a single TQC operation sheet. To calculate demand at capacity, divide the targeted monthly volume by the number of workdays in the month (see Figure 4.4).

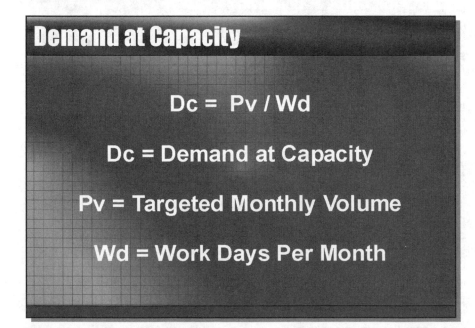

Demand at Capacity

$$Dc = Pv / Wd$$

$$Dc = \text{Demand at Capacity}$$

$$Pv = \text{Targeted Monthly Volume}$$

$$Wd = \text{Work Days Per Month}$$

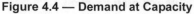

Figure 4.4 — Demand at Capacity

This equation provides the targeted number of units to be produced per day for demand at capacity. For example, if the designed monthly plan is based on 500 units and the total number of workdays per month is 20, the demand at capacity would be 25 units per day. Although the daily rate can be and is adjusted every day, flow lines are designed one time to accommodate demand at capacity volume. DFT is the only flow-manufacturing approach to design production lines in this manner.

Flow rates determine the design as well as the daily management of a flow process. They are based on actual daily units completed at the back of the flow process. In the flow-rate calculation, the use of effective work hours, or the amount of time that can be anticipated as actual work time, is required. As a typical example, production employees work a standard eight-hour day, with allowances for a 30-minute lunch and two 15-minute breaks. The remaining work time is then reduced 12 to 18 minutes a day to allow for quality discussions and personal time.

Based on this example, the total effective work hours would be 7.3. The flow line flow rate is equal to the specific daily rate divided by the effective work hours multiplied by the number of shifts per day (see Figure 4.5).

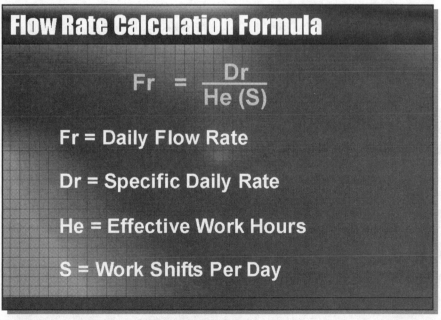

Flow Rate Calculation Formula

$$Fr = \frac{Dr}{He\,(S)}$$

Fr = Daily Flow Rate

Dr = Specific Daily Rate

He = Effective Work Hours

S = Work Shifts Per Day

Figure 4.5 — Flow Rate Calculation Formula

Thus, a daily rate of 50 units divided by 7.3 hours in a one-shift operation would yield a flow rate of 6.8 units per hour. If all other variables were equal, and the daily rate of 50 units were achieved from a plant operating with two shifts, the flow rate would be half that amount, or approximately 3.4 units per hour. The calculation would be 50 divided by 7.3 times 2, or 50 divided by 14.6. Flow rates are important in managing the production progress throughout the day, particularly in the high-volume manufacturing processes. The Flow Rate would normally be measured at the end of the production line.

Operational Cycle Time

"Operational cycle time" is based on demand at capacity. It is the targeted-work-content time required for a single person or machine to produce a single part or product within the flow process. The operational cycle time calculation establishes the "takt" (rhythm or beat) of a process. For a product produced using multiple processes defined in product synchronization, the takt time is calculated for each process. It is a calculated numeric-time value based on the targeted work content. Operational cycle time is the reciprocal relationship of the flow rate, as shown in Figure 4.5.

Simply stated, the operational cycle time equals the effective work hours in a shift multiplied by the number of shifts per day divided by the demand at capacity. With demand at capacity of 5 and 7.3 effective work hours per shift, one shift operation would have an operational cycle time or process takt of 1.46 hours per unit, or (as preferably stated in minutes) 87.6 minutes per unit.

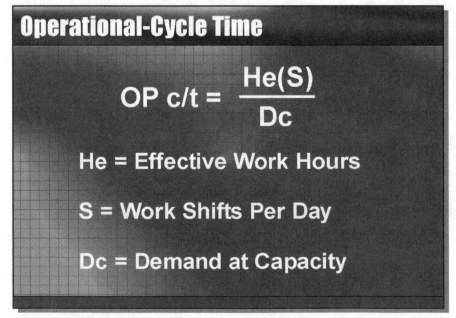

Operational-Cycle Time

$$OP\ c/t = \frac{He(S)}{Dc}$$

He = Effective Work Hours

S = Work Shifts Per Day

Dc = Demand at Capacity

Figure 4.6 — Operational Cycle Time

The operational cycle time formulas and calculations would be used for all flow-manufacturing lines, regardless of the product or volume. The targeted work content is identified based on the operational cycle time calculation. The TQC sequence of events is then grouped into pieces of work equal to this targeted-work content. These ideally equal, grouped pieces of work are defined as flow operations (see Figure 4.7).

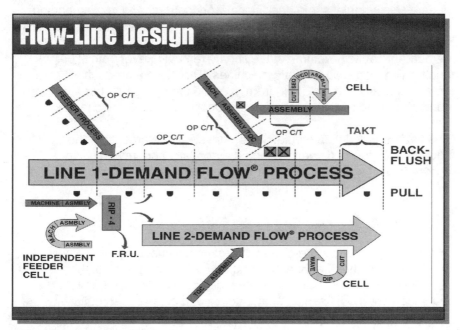

Figure 4.7 — Flow-Line Design

Targeted Rate, C/T Relationships

The higher the demand at capacity or the higher the volume required, the shorter the designed operational cycle time and the faster the process takt time. Likewise, the lower the demand at capacity or the lower the volume required, the longer the designed operational cycle time, and the slower the process takt time. These demand-at-capacity volumes for products to be produced determine the corresponding work content required to achieve the targeted rate or volume. Once the targeted work is defined, the TQC sequences of events are independently grouped into machine and labor operations. Each operation would ideally have actual work content equal to the targeted operational cycle time of the production line or cell.

Volume Adjustments: People

This designed operational work content and corresponding TQC quality criteria is now defined and fixed. To adjust the volume of products required to meet specific daily rates, the manufacturer either removes people from the operations and turns off machines or works fewer hours per day. Nevertheless, the operational work content and corresponding quality criteria do not change. For example, the actual volume of products produced may simply be reduced by 50 percent from the designed demand-at-capacity volume at the line by removing every other person and turning off the appropriate machines. The flexible production employees simply move (flex) from operation to operation, but the work content and the quality criteria at each operation would not change.

The flexible production employees are invaluable elements in the flow-manufacturing processes, and their certification, reward, and compensation should reflect their new responsibilities and contributions.

Line Design

In flow manufacturing, the design of production lines and cells always supports the highest required volume and the corresponding shortest required operational cycle time. When designing a flow line or cell, the manufacturer should seek management's and sales' most realistic prediction of the highest anticipated capacity volume for each product.

This required volume must look forward at least a year into the foreseeable future. The flow manufacturer then calculates the targeted operational cycle time based on this anticipated highest demand to calculate process takt times for the line. Next, the manufacturer designs a line with operational work and quality criteria equal to the corresponding process takt time. Line design in flow manufacturing also presupposes a flexible, multi-skilled workforce, which generally delivers higher productivity and client satisfaction.

Daily Volume Demand Changes, Line Does Not

As discussed earlier, changes to line layout aren't necessarily required with each change in demand-volume daily rate—as opposed to other approaches, such as TPS. The flexible employee in the flow process allows lines to run at lower demand volumes, with the removal of employees from required operations. A line or cell with fewer production employees than the total number of operations is known as a "line with a 'hole' in it."

Production employees simply flex from operation to operation to maintain the pull process. Employees removed from the line work on employee-involvement tasks, cross training, quality-improvement programs, or similar tasks until demand increases. The "holes" in the flow line move up and down the line as production employees flex to pull work to each operation.

Possible Imbalances

Once the targeted rates and corresponding takt times have been defined and the actual operational work content is established, imbalances between the target operational cycle time and the actual observed operational cycle time could occur. Labor-intensive operations can be adjusted by relocating material or work content. However, operations involving machines that effectively run at one speed require different techniques to adjust for the imbalance.

Adjusting for Line Imbalances

Flow manufacturing's objective is to balance actual work content with targeted operational cycle time. Refer to Figure 4.8 and consider a flow line where five successive operations have an actual work content as follows:

Operation 30 ....20.0 minutes

Operation 40 ....20.0 minutes

Operation 50 ....20.0 minutes

Operation 60 ....25.0 minutes

Operation 70 ....20.0 minutes

These operations are part of a flow line designed to produce 22 units per day. However, operation 60 is a machine operation that produces a unit every 25 minutes, no more, no less. Figure 4.8 illustrates the calculation of targeted operational cycle. The targeted process takt time of this line is 20 minutes, but the actual time to produce a part at operation 60 is 25 minutes. During the 7.3-hour day, the 20-minute operations would produce 22 parts, while the 25-minute operations would produce only about 17 parts. Since the line is targeting a volume of 22 units per day and the machine at operation 60 is capable of producing only 17 units per shift, the manufacturer would have three alternatives to correct this imbalance:

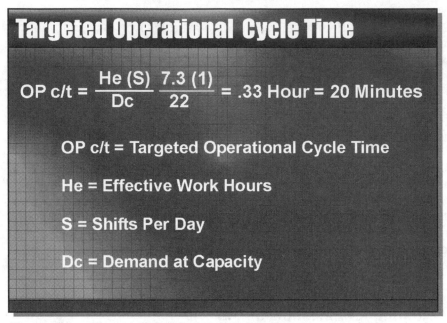

$$\text{OP c/t} = \frac{\text{He (S)}}{\text{Dc}} \frac{7.3 \, (1)}{22} = .33 \text{ Hour} = 20 \text{ Minutes}$$

OP c/t = Targeted Operational Cycle Time

He = Effective Work Hours

S = Shifts Per Day

Dc = Demand at Capacity

Figure 4.8 — Targeted Operational Cycle Time

1. Reduce the actual cycle time of the machine at Operation 60 to 20 minutes by eliminating any non-value-added time, such as setup or move time.

2. Obtain an additional machine capable of producing at least five units per shift.

3. Create an inventory of units around the machine that would allow the machine to run longer hours than the rest of the line.

The first alternative is always preferable. The second alternative is usually the most expensive, and the third alternative is the most frequent choice.

The number of inventory units required to allow the machine to work additional hours is calculated based on the imbalance between the actual time to produce a part and the targeted actual cycle time of the process.

During the 7.3-hour first shift, a buildup of five units between operation 50 and Operation 60 would result. The machine could work additional time on a second shift to process the buildup of five parts to operation 70 for the start of the next day. At the start of the next day, the inventory in the front of operation 60 would be zero, and the inventory in front of operation 70 would be five units. The inventory required to support this imbalance is referred to as an in-process kanban (see Figure 4.9).

The cost of another machine notwithstanding, the imbalance does not appear sufficient to warrant investing in new equipment.

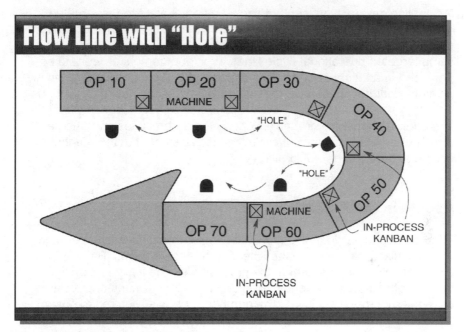

Figure 4.9 — Flow Line with "Hole"

Overtime or Additional Shifts

Based on the imbalance of units, additional hours of production are needed. They could range from adding a second shift to alternating operators and keeping the operation 60 machine running through lunch and breaks. An in-process kanban containing several units would exist before the machine and before operation 70. This in-process kanban would contain the units produced in overtime or on the second shift, and it would keep the line flowing and achieve the targeted daily rate.

If the imbalance resulted from a setup or other non-value-added work, that problem could be attacked vigorously. If the work content cannot be balanced, then an in-process kanban can correct the imbalance by providing a correctly sized point of supply.

Equalizing the Process

In flow-line design, work content should be equivalent to the targeted operational cycle time. Once this is understood, it is possible for an automobile manufacturer and an ordinary pencil manufacturer to have the same targeted operational cycle times. If eight automobiles were to be produced in an eight-hour day and eight pencils were to be produced in an eight-hour day, both have the same operational cycle time.

However, more people probably work on the automobiles than the pencils. The targeted operational cycle time is the same, but the number of people and machines would differ.

People in the Process

The operational cycle time defines the work content for each operation. This calculation also establishes the takt for each process. Barring a change in or improvement to the process after the initial line design, the operational work content is fixed and is no longer subject to volume changes (non-rate sensitive). The number of people required to support a process is derived from the labor time per unit. This calculation is very rate sensitive. Consider a process that has a daily rate of 25 units per shift, total labor hours from the TQC sequence of events of 36 hours per unit, and effective work hours in a shift of 7.3.

To calculate the number of people needed to support the process, multiply the specific product daily rate by its total labor hours per unit and then divide by the effective work hours in a shift, then multiply by the number of shifts (see Figure 4.10).

The process requires 124 people. If the line is not running at its designed capacity, the line has "holes." To fill these holes, people simply flex from operation to operation upon completion of the work content at their primary operation.

These techniques aid in designing and balancing a flow line. And once the daily rate and operational cycle time techniques have been mastered, along with an understanding of the pull techniques, the particular product or related technology used to produce it are irrelevant.

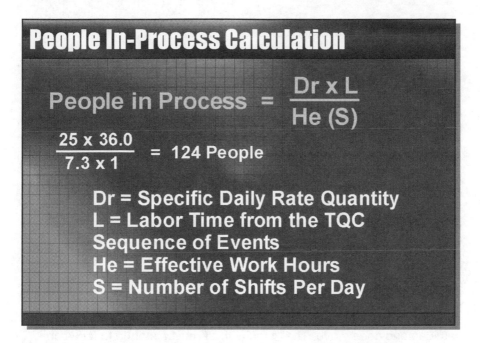

Figure 4.10 — People In-Process Calculation

Total Product Cycle Time

Total product cycle time (TP c/t) is the next key element of the production flow. TP c/t is the longest path of a flow process as measured from the completion of the product. This key value is the basis for the inventory investment required by the process.

It is also the basis for the absorption of overhead in the flow-manufacturing financial system. Along the TP c/t path, process improvements are listed in order of priority to eliminate non-value-added steps. TP c/t is a fixed number that is not rate-sensitive and does not change as long as the process remains stable. The elimination of non-value-added steps along the TP c/t path causes the path to shift location and continually refocus for ongoing process-improvement activities. TP c/t is calculated by timing the work content through the longest path of the product-development process.

Starting from the End

In a flow-manufacturing pull process, the daily rate emerges at the completion of the flow process. The last operation pulls from the previous operation, continuing all the way upstream to the calculated origin of the product. In the TP c/t calculation, the end of the line is always the starting point of the measurement (see Figure 4.11). Starting from the end working up the flow line to the calculated beginning, the TP c/t path can be determined by taking the longest step at each of the many decision points in the process. Each of those steps requires an analysis of which path (and its upstream feeder paths) is the longest. Beginning at shipping, going upstream through the production process, and moving off to feeder and other side or main processes, the longest path can be determined. TP c/t is the longest cumulative single path back through the process, regardless of whether it follows the main line or trails off to a feeder process.

After determining total product cycle time, the manufacturer can take steps to shorten it. With ongoing improvements to the process, the TP c/t path can and does move. The manufacturer must continue to focus on the total product cycle time, and eliminate non-value-added steps, such as setups and move time. TP c/t does not change without improvements to the production flow process.

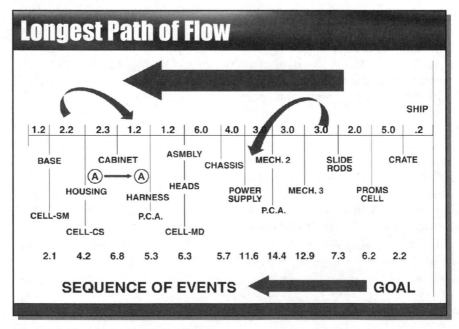

Figure 4.11 — Longest Path of Flow

Analysis of Flow Path

Analysis of the flow path (total product cycle time) always begins at the completion, or end, of the process. The analysis involves taking the work content, adding back to front at the point where the first feeder is consumed or required for final assembly. Calculating the work-content time from the back of the process to the point where the feeder is consumed—and adding the work-content time of the feeder process—yields the time through the first feeder. The computation continues from the back to the next point, where a second feeder process is consumed. Work-content time from the back of the process to the point where this feeder is consumed, plus the work-content time of this feeder process, is added and compared to the work-content time that was calculated for the previous feeder. The feeder process associated with the lowest number is eliminated as the analysis continues to each point where a feeder process is consumed. The goal of the search is to find the longest path (as measured in time). This is the total product cycle time required by the flow of the process.

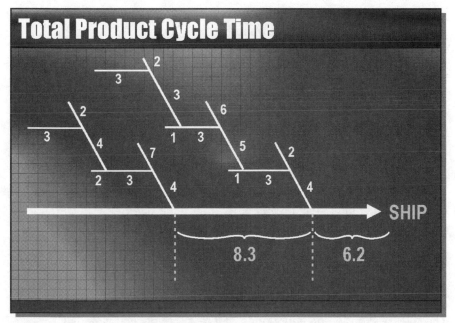

Figure 4.12 — Total Product Cycle Time

The Importance of Product Cycle Time in Demand Flow

Total product cycle time is of crucial importance in the Demand Flow process. It impacts inventory investment, absorption of overhead, response to customers, and process improvement.

Dictates Inventory Investment

First, total product cycle time dictates the minimum inventory investment required to support the process. The shorter the total product cycle time, the shorter the time that inventory must be on hand in the production process. In traditional subassembly manufacturing, in-process inventory is maintained for the lead-time days, weeks, or months required to schedule, queue, kit, and build each level of the multilevel product. In Demand Flow manufacturing, the product can progress through the flow process in less than the total work-content hours to build the product. As the total product cycle time shrinks, so too does the corresponding in-process inventory investment.

Basis for Absorption of Overhead

Second, total product cycle time is crucial because it is the basis for the absorption of overhead. The efficient Demand Flow manufacturer does not apply overhead based on labor, because labor is not a primary focus and it represents the smallest and shrinking portion of product cost. Total product cycle time is a consistent and fixed basis for applying overhead. As total product cycle time decreases, overhead is not fully absorbed. Management and sales face increasing pressure to add new products and increase sales volume to support the same overhead. Traditionally, under-absorption of overhead negatively affects fiscal standing. It can indicate an insufficient number of labor hours (to meet the budget). This can trigger an increase in inventory to absorb the overhead. In Demand Flow manufacturing, the under-absorbed overhead can be a powerful management tool to force process improvements because of reduced total product cycle time.

Quantifies Optimum Response Time

The third reason for the importance of total product cycle time is its ability to quantify the optimum response time that can be achieved from the overall production flow line. In the case where the flow line is just the final-assembly process, TP c/t represents the response time to the customer.

Where the TP c/t is calculated by the addition of the final-assembly processes and the upstream-feeder processes, machine cells, and other factors, then customer response may require less time than the total product cycle time, due to strategic use of in-process inventory in the overall manufacturing environment. We will discuss this dynamic later in this book.

Guides Process Improvement
The fourth reason for the importance of total product cycle time is its ability to guide a process-improvement program. Management should emphasize the high priority of process improvement/employee involvement throughout the TP c/t path of the process. The world-class Demand Flow manufacturer must strive to decrease the number of non-value-added steps to reduce inventory-investment time and total product cycle time, along with the corresponding absorption of overhead.

Increasing Inventory Turns

Total product cycle time dictates the in-process inventory investment in Demand Flow. Reducing total product cycle time is a primary objective. Total product cycle time is determined by the work content along the longest path of the process, which is usually a shorter period of time than the total work content—or the total amount of time it takes to build a product. For example, in a dramatically simplified line layout, building a product may require 20 total hours. Those 20 hours may include an eight-hour feeder process consumed toward the end of the line. If the consuming line is a totally sequential process of one step after another for 12 hours, the longest path could be 12 hours, assuming the eight-hour feeder process occurred simultaneously. Even though building the product takes 20 total hours, parts and raw materials—all inventory—need only be in the process for 12 work hours. Shortening the total product cycle time to reduce the time inventory must be present—and increasing the material turnover—are primary objectives from a financial standpoint.

In a demand-driven world, single-digit inventory turns are unacceptable, and they can cripple a manufacturer's competitive power. Traditional methods utilizing schedule-based systems (such as MRP II) produce traditional results. Competitive goals—such as achieving 20-plus inventory turns annually—require flow-manufacturing techniques and systems.

Developing a Competitive Edge

DFT is the foundation for world-class manufacturing. Demand-driven business strategies take advantage of the resulting improvements in quality and customer response times to dominate their respective markets and industries. TQC sequence of events, total product cycle time, operational cycle time, and TQC operational method sheets are the prerequisites to developing a flow-production process. These techniques and formulas become the basis for a Demand Flow production process, which begins with identifying the natural product flow, work content, corresponding quality criteria, and non-value-added steps. Then, the operational cycle time is calculated based on the highest required rate.

Operational method sheets graphically identify work content and quality criteria based on the targeted operational cycle time. Total product cycle time is then calculated as a guide to inventory investment, overhead absorption, customer response, and process improvement. By identifying these essential building blocks of a Demand Flow process, the world-class manufacturer has the framework for a powerful, competitive tool. With increasing market pressure from powerful new competitors—in addition to the shortened product life cycles—these techniques become essential to industry leaders where speed to market dictates profit and market share.

Chapter 5
Customer Responsiveness
and Linear Planning

Product acceptance and sales dictate survival and success. Without sales, even the best products and the most compelling goals become mere dreams. Competitive markets demand that successful corporations become increasingly responsive to changes in customer demand. Immediate response to fluctuations in product volumes and model mixes are key elements of a Demand Flow process.

Linear planning and material forecasting determine the production volumes required to satisfy the quantity and timing of anticipated demand. They also enable the transition from lumpy, infrequent and inaccurate forecasts to linear, daily production plans. The Demand Flow approach calls for flexibility, total product cycle time, forecast consumption techniques, demand-time fences, and planning-flexibility fences. These tools also draw together historically antagonistic constituents in manufacturing companies. These antagonists—sales, materials, production, and product planning—work on a much more cooperative basis in Demand Flow manufacturing, developing better forecasts and plans in a more manageable system.

Traditionally Long Lead Times

The goal of the world-class manufacturer is to deliver high-quality products to the customer on a timely basis while minimizing inventory investment and keeping the production process as linear and as flexible as possible. Traditional scheduled manufacturing has long, fixed lead times for purchased material. In the 1990s, procurement teams focused on reducing these lead times, often with excellent results. Nevertheless, adversarial relationships have continued between manufacturers and their suppliers.

While many companies have eased these conflicts, the manufacturer often found that progress toward reducing lead times slowed as variation in demand forecasts increased. This situation created an impasse, leading manufacturers to refocus their efforts on the bigger picture of the supply chain. Supply chain management applications bring more force to bear on the lead-time issue and the resulting service problem. In traditional manufacturing, long lead times force suppliers to procure adequate material to produce their own products to meet the consuming manufacturer's demand. Thus, the supplier's lead time—which includes scheduling, queue, and production time—must be added to the consuming manufacturer's lead time—which also includes scheduling, queue, and production time—before the ultimate customer can be served (see Figure 5.1).

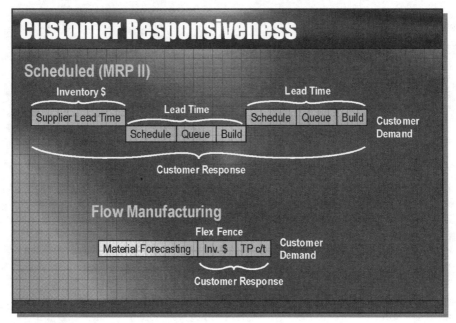

Figure 5.1 — Customer Responsiveness

Targeting Customer Responsiveness

In linear Demand Flow manufacturing, a long-term purchasing/planning forecast is shared with suppliers. This forecast allows the supplier to plan material and production processes. In exchange for this longer-term forecast, the supplier reduces fixed lead times to the consuming Demand Flow manufacturer. Thus, the supplier's lead time is synchronized with the very short production time needed to make the supplied part.

The supplier's short production time, combined with the Demand Flow manufacturer's total product cycle time, becomes the manufacturing lead time. The number of suppliers per manufacturer goes down because mutually beneficial partnerships replace informal business relationships.

The supplier gains up to 100 percent of the available business, while the manufacturer benefits from consistently high-quality products supplied within flexible, shorter lead times. With the very long lead times of traditional scheduled manufacturing, the customer waits a long time to receive product, or the manufacturer buffers its forecasting errors with additional inventory. The inventory is in the form of finished goods, purchased parts, or work-in-process (often in a combination of all three). As total product cycle times and flexible lead times with suppliers shrink, the manufacturer evolves from a conventional finished-goods business model to a make-to-order business model. Inventory levels decrease, and customer responsiveness and service increase.

In a traditional scheduled-manufacturing environment, unpredictable spikes in product demand often occur. Production ramps up to a much higher volume in a short period of time in order to meet the demand spikes. Schedules increase to reflect the higher quantities. These demand spikes often cause chaos on the production floor, with continuous schedule revisions required to accommodate the seemingly arbitrary demand pattern. In contrast, Demand Flow manufacturing practitioners strive to achieve revenue and forecast plans in a customer-responsive and controlled ("spikeless") fashion. Daily production rates gradually increase or decrease to follow customer demand and meet plan targets.

Small Changes Allowed

In a Demand Flow manufacturing environment, sales departments can make frequent small changes in a short period of time. The goal for all corporate departments—sales, materials, production planning, and production—is to meet market-driven demand while minimizing inventory investment and keeping the production process as linear as possible. The result of the consistent blending of these initiatives is a very powerful manufacturing process facilitated by the flexible demand-based production plan. Elements of the master plan include forecast consumption techniques, demand-time fences, planning-flex fences, actual orders, production capacity, finished-goods management, the product-option mix, and the products available to promise to the customer.

Demand Flow Is Market Driven

For many people, the term "accurate forecast" is something of an oxymoron. For these cynics, forecasts fall into one of two categories: lucky and lousy. Forecasts that don't result in actual customer orders result in burdensome inventory, either as components or finished goods. Understated forecasts eventually turn into shortages—if not for a particular order pulled up then for orders to be pulled in the future.

The equation below shows the typical relationship of the number of orders shipped early due to a pull-up and the number of orders shipped late due to a pull-up, where "L" represents orders shipped late and "P" represents orders pulled up:

$$L = P^2$$

In defense of sales, it is a department bedeviled by customers who sometimes do not know how much product they require. When they do know, they sometimes keep that information from the sales department. Additionally, manufacturing organizations and production processes can be unresponsive, contributing to poor customer satisfaction levels and hampering sales. The key to Demand Flow manufacturing is for manufacturing and sales departments to collaborate in negotiating a requirements plan for a product that can be manufactured linearly. This plan also must meet customer needs. The demand-based production-planning process helps companies reach this consensus.

The Demand Flow production process is designed for a capacity volume, which corresponds to the lowest required operational cycle time. Typically, small changes in a fluctuating sales plan don't necessitate changes to line layout or production processes. If the changes become larger than the production process can support (more than 30 percent above planned capacity or less than 50 percent of the designed volume), then the planning process can help determine how to make the changes in the flow line. Otherwise, sales and manufacturing must renegotiate the plan.

Sales and Manufacturing Alliance

Customer and supplier involvement is paramount to developing a flexible manufacturing plan for a company using DFT. Continually changing sales forecasts require a flexible demand-based plan. On the one hand, the flow-manufacturing process is market driven. On the other hand, the flow-manufacturing process must keep the sales forecasts within reasonable limits. Sales assumes responsibility for finished-goods inventory—not physically, but in terms of inventory management and accountability.

The reason is that in the Demand Flow process, any finished-goods inventory is usually the result of an order from sales that did not originate from a customer order. Sales must predict what the customer will want and assist manufacturing in satisfying that demand. Manufacturing must press for forecast accuracy within certain parameters; but more significantly, it must assist in the effort. When the product demand from customers can only be met with finished-goods inventory, then the sales and manufacturing teams must structure a demand-based plan that sets the level of inventory according to demand and company policy.

Sales Preference for Demand Flow Technology

Management and sales personnel tend to prefer DFT for three reasons. First, the product moves through production more quickly. Second, Demand Flow manufacturing creates a very flexible production process, which accommodates many small changes to the production plan in a very short period of time. These are not major changes but multiple smaller adjustments. No manufacturing process, including Demand Flow manufacturing, can go up or down as fast as sales would like, but the greater flexibility of Demand Flow manufacturing provides a significant sales advantage. Traditional manufacturing companies tend to freeze the production plan for a month, a quarter, and, in some cases, even longer. This ties the hands of sales, restricting their ability to make more sales to customers or to manipulate finished goods in a desired or opportunistic fashion. Finally, the third reason management and sales prefer DFT is that the Demand Flow process produces high-quality products in the manner and time intended. In traditional manufacturing, companies tend to meet a majority of the monthly or quarterly goals at the end of the production month or quarter. Therefore, customers who requested product at the initial and middle stages of the month have little chance of obtaining their requested product delivery.

Total Customer Satisfaction

Companies traditionally review customer service measures on a weekly basis. Thus, if a product ships on the last day of the week (Friday) when it was actually required to ship on the previous Tuesday, it is considered by the manufacturer as an on-time delivery, even though the customer considers the shipment to be late. In Demand Flow manufacturing, the efficiency of making the daily rate through linear manufacturing gets product to customers when promised, as promised. It is an extremely predictable process, which means that instead of explaining quality problems or shipping delays to the customer, sales can concentrate on new and expanded business.

Sales not only has the ability to make small changes in a short period of time to meet customer demand, they can also sell higher-quality products due to the inherent quality that is built into the TQC production process. Sales also gains the advantage of faster and timelier delivery. From a manufacturing standpoint, extreme peaks and valleys in demand disappear. Manufacturing cannot and should not accelerate from two units one day to ten units the next day. But production can go from two units per day to three units and from three units to four units, until reaching the desired volume over a controlled period of time.

Quicker Response

Traditional scheduled manufacturing may take six to ten weeks in accumulated lead time to deliver a product that Demand Flow manufacturing techniques can deliver in a matter of days. The major difference is the elimination of lead time and other traditional tools such as scheduling, planning, and queuing at every operation. Demand Flow manufacturing does not recognize traditional manufacturing lead-time techniques. In Demand Flow manufacturing, suppliers use a long-term planning forecast only to assist in their planning visibility. That planning forecast has tolerances built in to accommodate changes in demand. These planning forecasts are available long before the release of a purchase order under conventional manufacturing.

This gives the supplier a far greater jump on meeting the Demand Flow manufacturer's requirements, without the need for long lead times or the usual delays in meeting the targeted production date.

In a demand-based planning process, sales and manufacturing negotiate a demand-based production rate for each product. The corresponding output of the production process is then calculated for the product or group of products. A flexible range of volumes is negotiated with suppliers for delivery of supporting material. Only then can a company synchronize sales' negotiated volume, the daily production volume, and flexibility with the suppliers in a consistent manner. Within a very short period of time (no less than the manufacturing total product cycle time), no changes are made to the production plan. Outside this short fixed interval (TP c/t), production, sales, and eventually suppliers can negotiate planning flexibility.

Three Masters

A consistent strategy should govern discussions related to pricing, market share, costs, and product volume. This strategy is based on the synchronization of the three masters of production: sales, suppliers, and the manufacturing process itself. Sales must not promise what suppliers and their production process cannot deliver in either quantity or time. Production should not deliver goods that sales cannot sell. Suppliers should not deliver materials too soon or too late to serve sales. Coordination and synchronization of the three masters occurs through the art of negotiation.

Demand Flow Technology links the three masters inseparably in a methodology referred to as demand-based management.

Initial Determinations for the Plan

Certain information and decisions precede demand-based planning. One such piece of information is the manufacturing calendar—the days to be worked, the number of shifts, the effective work hours, and the annual holidays. The product must also be analyzed with an eye toward delivery policy considerations in the demand-based production plan.

Is this a product that the market is willing to wait for, such as a custom automobile or other major product with a wide assortment of options that is built to customer specifications? If the market is willing to wait, how long will it wait? If it is a product that the market won't wait for—such as a toaster or a telephone—the product must be readily available, or the customer immediately turns to a competitor. If the plan dictates finished goods, building to stock and monitoring finished-goods inventory for quantity and option mix must be considered. With build-to-stock products, manufacturing process treats finished-goods replenishment orders the same as a customer order. However, the final determination of whether units should be shipped to finished goods or to a customer—or whether to shut down the production line—is largely a function of management and sales. As the Demand Flow manufacturing total product-cycle time continues to decrease, a transition toward a make-to-order manufacturing environment occurs. The manufacturer can than reduce or eliminate finished-goods inventory.

Under-Capacity Planning

A demand-based planning environment offers several advantages to planning at a level under manufacturing capacity:

- Maintaining quality in a process that runs at full capacity is difficult. In the TQC process, employees utilize total quality control techniques to perform quality-verification steps, in addition to the usual production work. Quality is the primary objective, and it is never compromised in a TQC process. Verification and TQC are as high a priority as the work itself.

- The production process becomes linear, and the daily rate (no more and no less) can be achieved every day.

- Running at under-capacity allows for employee training and involvement. Employees have the time to learn additional operations and train for even more flexibility.

- Other advantages include more time for machine and equipment maintenance, process improvements, plant cleanup, and meetings on employee involvement and quality.

- If the daily rate comprises approximately 7.0 or 7.5 hours for an 8-hour shift, the balance of the shift should be used for the purposes outlined above. If the daily rate is met in substantially less time (6.0 or 6.5 hours into the shift, for example), the process involves too many people, or the calculation of requirement and/or actual capacity may be inaccurate. Both cases require review and subsequent adjustment. Of course, if necessary, the Demand Flow process can run at 100 percent capacity for short periods, but that capacity should be discouraged for lengthy periods.

Demand-Based Planning

While traditional manufacturing usually creates production plans for an entire month and then abandons the monthly plan in favor of a weekly schedule, Demand Flow manufacturing executes a demand-based plan for each day. Demand Flow manufacturing's demand-based plan is a specific, calculable number with three outputs:

1. The required daily rate

2. The targeted flow rate

3. The number of people and/or hours of machine time in each process

If any two elements are known, the other can be calculated. The information can determine the corresponding flow rate of a process, the number of people needed to support a specific daily rate, and the maximum daily rate that a process can support. If one of the elements becomes variable, the impact on the other two variables can be analyzed. For example, if the required rate goes up, the additional people needed in the process to support the higher rate can be calculated, as well as the flow rate needed to reach the higher required daily rate. In the Demand Flow manufacturing process, the observed flow rate of units coming out of the back of the process is managed and charted.

Tracking Shorter Periods

The Demand Flow process reaches the point of completing a unit at the targeted flow rate. Management reporting evolves, for example, from tracking 420 units per month down to completing one TQC unit every 21.9 minutes. The inability to achieve the daily rate points to a problem in the design or balance of the process.

Forecast Control Chart

The demand-based planning process begins with sessions between production and sales to negotiate a production volume for each product in the process. This negotiated volume can and will change over time, but it initially defines the base level—a critical number for the application of demand and planning-flex fences. Some companies exclusively build products to order. They don't worry about a forecast from sales because of their make-to-order environment. This situation is rare. Other make-to-stock companies don't use actual orders from sales. Instead, they build strictly to a forecast. They are not concerned with actual orders from the customer. Most companies require a forecast, at least for the material-planning process. Most manufacturers blend their forecast and the actual orders in their process—primarily manufacturing to fill customer orders but buying material according to their forecast.

Sales' Call

"Forecast consumption" refers to the blending of the forecast and actual orders in a manner that is acceptable to both manufacturing and sales. The premise behind forecast consumption is that sales has a specific period of time to book or commit a customer order into the manufacturing process.

If they miss that time (referred to as the "demand-time fence"), then they lose any forecasted quantities for the product that are greater than the actual customer orders at that particular point. If sales still believes a customer order will arrive at the last minute but just outside the demand-time fence, sales can place a finished-goods replenishment order. The products associated with the finished-goods replenishment order go into a finished-goods area and become "the property" of sales.

If the customer order does come in, then the product can ship on time. If not, sales becomes "owner" of the inventory until they can sell it. It becomes sales' call at that point—just outside the demand-time fence—whether that inventory risk is appropriate and/or whether the customer commitment will come for the order. The length of the demand-time fence typically falls between the total product cycle time and the product lead time quoted to a customer. The duration of the demand-time fence also may include administrative response time and paperwork-processing time.

Forecast/Orders to Actual Orders

If, for example, a commitment to the customer specifies that the time between placing and shipping the order is three weeks, then the demand-time fence must be three weeks or less. Sales then has until the third week to consume or book orders against the forecast. Once inside that fence, manufacturing builds strictly to orders already received from the customer or for finished-goods replenishment orders (based on sales' high degree of confidence in their projection to turn the forecast into actual customer orders). The formal planning system should automatically cut back to actual orders inside the demand-time fence. But if the order has not been received at that point, it is assumed that the order will not be received.

Thus, according to the definition of the demand-time fence, only actual customer and finished-goods replenishment orders affect the calculation of total demand. Outside the demand-time fence, the greater of either forecasted orders or actual orders constitutes total demand. In most cases, forecasted orders are the greater amount. If the actual orders outside the demand-time fence are greater than the forecasted quantity, then the actual order quantity is used for total demand, with recognition that sales has overbooked the forecast for that particular period, requiring adjustments to the forecast.

Total Demand

The definition of total demand inside the demand-time fence is the sum of actual customer and finished-goods replenishment orders. Total demand outside the demand-time fence is the greater of the forecast or actual customer orders. Therefore, the first step in calculating the production plan is to determine the total demand based on the forecast consumption technique. The forecast consumption technique is applied against the forecast and actual orders to determine total demand. For example, with a forecast for 100 units per week in the fifth week and actual orders of 105 units per week in the same week, the total demand does not pick up the forecast of the 100 but instead picks up the actual customer orders of 105—resulting in a total demand of 105 units (see Figure 5.2). In the first week, which is inside the demand-time fence of one week, actual customer orders may be for 100 units, with a forecast for 120 units. In this scenario, the total demand in the first week will be for 100 units, unless sales decides to give production a finished-goods sales order for up to 20 units for the balance of the forecasted units.

Demand-Time Fence Governs

Unless sales decides to buy the unsold units for finished goods, the 120-unit forecast is overridden, since it is inside the demand-time fence and only the 100 units actually booked as customer orders are acknowledged. In the third week, which is now outside the demand-time fence, a forecast of 120 units with actual customer orders of 100 units is treated differently.

The 120 forecasted units are treated as total demand, since the forecast is outside the demand-time fence and is greater than actual orders. The total demand calculation is the basis of demand- and planning-time fences. In developing the master production plan, the total-demand control chart that is fixed only for a short time blends both forecasts and orders. The chart then changes to reflect the flexibility parameters of material and capacity. A demand-time fence, planning-flex fences, linearity, and mixed-model sequencing considerations are crucial to master-plan development. It is critically important to negotiate with sales for quantities that suppliers can support. Allowing 20 percent flexibility to sales while getting only 10 percent flexibility from suppliers is a shortcut to disaster.

Likewise 20 units can't be promised to sales when only five are in process for the same period of time. Parameters developed represent limits that, if exceeded, must be addressed in another way.

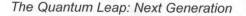

Calculating Total Demand

	WEEK 1	WEEK 2	WEEK 3	WEEK 4	WEEK 5	WEEK 6
Forecast	120	105	120	135	100	120
Actual Orders	100	100	100	115	105	75
Total Demand	100	105	120	135	105	120

TP c/t

Figure 5.2 — Calculating Total Demand

Establishing the Master Plan

Initially, a master production plan includes a forecast, actual orders, and what remains as available to promise (ATP). ATP is the difference between the total demand and the actual customer orders booked into a particular planning period. If ATP is a positive quantity, sales can still commit the positive quantity of products to customers for delivery in that period of the master production plan. ATP inside the demand-time fence for a true build-to-order manufacturer is zero. However, when sales decides to place an uncommitted finished-goods order inside the demand-time fence, then ATP is positive inside the fence until the finished-goods order is committed to the received customer order. Similarly uncommitted finished-goods inventory is also ATP for immediate shipment. Outside the demand-time fence, ATP is the total demand minus the actual customer orders that have been committed. Whether they are actual customer orders or committed orders by sales via finished-goods replenishment is irrelevant to manufacturing. The ATP figure serves sales and customer service as a guide to when they can promise product. Sales or customer service cannot commit a customer order either inside the demand-time fence or in a period when ATP is zero. Initially, the forecast itself determines the production plan, then a blend of forecast and actual orders, then actual customer and finished-goods replenishment orders alone.

Sales either receives sufficient orders from customers, reduces forecasts to actual orders, or issues an order of replenishment to finished-goods inventory. After the smoke of the initial forecast has cleared, the grace period ends, and the reality of what has transpired takes control. Material inventory may be delivered based on the total demand and therefore needs meticulous management within the acceptable control limits.

Time Fences

A demand-time fence allows no changes or deviations whatsoever inside its time horizon. Production rates are fixed for this short period of time, which is at least as long as the total product cycle time. The time horizon set by the demand-time fence includes the total product cycle time plus some short time period for work released called "administrative response." Thus, within the demand-time fence, the production plan does not change.

Therefore, under a negotiated production plan of 100 units per day, production is limited to precisely 100 units per day—no more, no less—within the demand-time fence. Sales looks to make required changes to the production plan outside the horizon set by the demand-time fence. As the total product cycle time decreases, and with it the demand-time fence, sales can make changes in a shorter period.

The shortening of the total product-cycle time helps achieve a key element in the manufacturing process: flexibility. Controlled flexibility is the primary objective. For example, if total product cycle time for a particular product is three days, the daily rate does not change within the three days because products are either in the actual production process (TP c/t) or within the demand-time fence horizon. As a result, by producing 100 units per day with total product cycle time on the third day, a manufacturer produces 100 units each of the three days. If total product cycle time can be reduced to two days, the demand-time fence can be moved in from the set horizon by one day. Demand-time-fence and planning-flex-fence techniques are always applied against the total-demand quantity previously computed. If the forecast had been greater than actual customer orders, and sales wanted the forecast quantity produced in anticipation of orders, then sales could issue a finished-goods replenishment order outside the demand-time fence. This additional quantity—thus authorized— progresses inside the demand-time fence and becomes part of the stable production plan. Outside the demand-time fence, sales can make changes to the total demand. After the short, fixed period, a longer, gradually increasing and more flexible period follows.

Planning-Flex Fence

This flexible period, which gives sales substantially greater flexibility in responding to customer demand, results from the planning-flex fence. It is critical that planning-flex fences negotiated with sales synchronize with the production process and suppliers. Planning-flex fences allow plus and minus percentage changes to total demand within a particular time period. Typically, the recommended number and magnitude of flex fences is three. For example:

1. 5 percent flexibility within two weeks

2. 10 percent flexibility within four weeks

3. 15 percent flexibility within six weeks

Since the Demand Flow production process is designed at the highest targeted capacity to accommodate market fluctuations, it is reasonable to achieve a 10 to 15 percent increase in a production plan by working additional hours and without requiring a line design change (see Figure 5.3). Typically, the greatest planning-flex fence should not exceed the highest rate for which the production process is designed. Any change beyond this capacity may require a major redesign of the production process and the addition of people and machines.

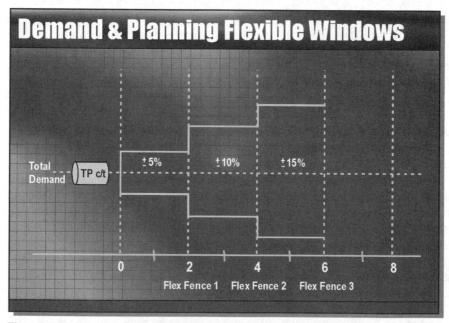

Figure 5.3 — Demand and Planning Flexible Windows

Level and Limit of Flexibility

The combination of fences provides both a level and limit of flexibility over time. The level and limit are achieved through a negotiated balance of the three factors:

1. Sales requirements

2. Suppliers' capability

3. Total product cycle time

Contracts with flow suppliers must reflect the necessary flexibility. The conversation, coordination, and negotiation between sales and production result from planning meetings conducted at least once a week. Led by sales' tracking, the two departments work to avoid spikes and moderate increasing and decreasing sales trends. Increasing demand progresses inside the demand-time fence and results in increases to the daily rate volume manufactured. Similarly if demand decreases, then the daily rate decreases in time as the demand quantities mature within the demand-time fence to become the daily rate. The objective is to follow the customer demand directly and manage substantially increased or decreased sales without creating production line upheavals (see Figure 5.4).

Tracking, reporting, and negotiating any of sales' violations are elementary to the success of the planning process. The negotiated flexibility among the three factors must be consistent, and product linearity must be achieved.

Demand-Based Plan

	WEEK 1	WEEK 2	WEEK 3	WEEK 4	WEEK 5	WEEK 6	WEEK 7	WEEK 8
Production Plan								
Forecast								
Actual Orders								
Total Demand	110	110	120	110	100	125	130	100
Production Plan	110	110	115	115	110	115	120	110
Production Plan Against Capacity								
Forecast								
Actual Orders								
Total Demand								
Production Plan	110	110	115	115	110	115	120	110
Capacity	130	130	130	130	130	130	130	130
% Capacity	85	85	88	85	85	88	92	85

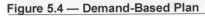
Figure 5.4 — Demand-Based Plan

Demand Flow Materials Management

Once the total demand is known, the flexible percentage (amplitude) in each time fence is applied against the current daily-rate quantity to calculate the high and low values of each flex fence. The objective is to define a tolerance range for each flex-fence time period (range) compared to the total demand for each period that falls within the range of that flex fence. The result is a series of master-plan quantities that mature with changes in the demand-time fence. The changes are not cumulative but are always applied against the negotiated production plan volume of the product. The master production planning process should be verified in a simulation mode. Any reported violations are negotiated, and the master production plan changes accordingly. This requires a simulation mode in the formal computer planning system to calculate total demand and apply flex-fence percentages against it—without driving down to the parts level and giving purchasing requirements to buy material against an unapproved plan.

Internal negotiations should take into account that the key week in the production planning process is the first week outside the demand-time fence. Once the move has been made inside the demand-time fence, the production plan is fixed and the agreed-on plan is executed. While in the first week outside the demand-time fence, sales and manufacturing must agree on the plan that moves into the demand-time fence. At that point, one of two things can happen:

1. The total demand must be changed to match the negotiated level, or

2. A new level must be negotiated.

If a new level is negotiated, then demand and planning-flex-fence percentages are applied against the new level for all periods of the planning horizon.

Flexibility Gains Market Share

Using the time fences, a likely scenario for the flow of materials typically follows negotiations between suppliers, sales, and the production process: production manufactures 100 units per day for the first week (a period sufficient to cover the demand-time fence, which includes the total product cycle time).

Supplier parts sufficient for that production volume are available. Production proceeds with no deviation through this initial period within the demand-time fence.

For the following two weeks, capacity has plus or minus five percent flexibility. Sales could order anywhere between 95 and 105 units per day but no more or no less. Then, in a later flex fence (perhaps four or eight weeks later), the flexibility increases to 10 percent and, after a similar period, reaches 15 percent. At that point, sales could order anywhere between 85 and 115 units per day but no more or no less. The gradually growing range of flexibility coordinates and takes into account manufacturing and supplier capabilities, balanced against sales' demands. Contracts with suppliers must specify similar flex, or inventory must be carried for long-lead-time items to meet flex requirements for non-contracted items. These flex-fence percentages (amplitude) can vary for different products. Product-demand patterns and each product's sales strategy assist in determining the actual percentages for flex fences for each product. These desired flex amplitudes are cumulated for products produced on the same flow line, with capacity constraints designed into the line via the determined demand at capacity. As expected, the higher the demand volume of a product, the more linear the demand pattern and the less flex amplitude required for a product. Low-volume products tend to have wider deviations in demand and thus merit higher flex percentages.

The manufacturer can accommodate higher flex percentages because the products do not affect capacity as dramatically as the high-volume demand products. The sales and manufacturing teams must achieve a balance that allows for setting flex amplitudes that suit the market strategy and that work within the available capacity constraints. Here, the previous discussion of designing the flow line to run under capacity serves the objective of a linear plan result.

Developing a Master Plan

There are eight basic steps to developing the master production plan, which evolves largely through weekly negotiations between sales, production, and production planning.

1. The eight-step process begins with receipt of a forecast and actual orders from sales.

2. The forecast and actual orders received via forecast consumption techniques determine total demand, period by period.

3. Once the total demand has been determined, demand and planning-flex-fence percentages are compared to the total demand. Any violations are reported to sales, which is responsible for appropriate changes to the forecast or actual orders, depending on where the violations occurred.

4. Once sales has made the appropriate changes, the master production plan can again be calculated to arrive at a total demand without violation. Initially, many violations are to be expected; but once the process is understood and the needed changes made, the number of violations should diminish.

5. Once an acceptable forecast has been received, a production plan is developed.

6. Another check ensures that production capacity can support the production plan. An under-capacity plan of about 85 to 92 percent offers the advantages of fewer quality compromises, a greater certainty of making the daily rate, and more opportunity for training.

7. The plan can then be broken down into a daily rate using a manufacturing calendar, which provides the number of workdays in a week, upcoming holidays, and so forth.

8. The final step is to consider product options that are built and that determine the mixed-model sequencing.

Negotiating Violations

Moving goods to customers or to finished-goods inventory can eliminate some violations that exceed the outer limits of the flex fences. Customer service may call a customer and ask that an order be taken a few days early or, to address the opposite problem, a few days late. The key here is to perform these negotiations in a proactive time frame and thus avoid last-minute changes.

The process then moves from an "initial" forecast to a "negotiated" forecast. The rescheduling of actual orders (with customer cooperation) and moving products to finished-goods inventory are important tools for sales and customer service in achieving compliance. This is another reason sales assumes ownership of finished-goods inventory.

Production Plan/Capacity Check

Developing the production plan is next, and leveling the peaks and valleys with linearity is a primary objective. The varying production totals of 120 units per week and 110 units per week average out to a rate of 115 per week, and a double check ensures that volume and materials are adequate for the plan.

A further check involves analyzing the volume against planned capacity to determine a percentage of capacity. A production plan running 80 to 89 percent of planned capacity is desirable (see Figure 5.4). Developing a production plan that is different from a negotiated total demand is an optional and prudent step. Manufacturing must agree that it can support the plan. If sales also agrees, the peaks and valleys can be leveled. This isn't always possible, especially in cases where the leveling may negatively affect customers. This "smoothing" process allows better linearity and consistent production operations.

Daily Rate/Mixed-Model Sequencing

With five workdays per week, the 115-unit weekly production plan translates into a daily rate of 23. At this point, a check of required options verifies that machines, people, and cycle times are sufficiently similar so that the different option units can run on the same line or cell by using group-technology techniques.

When different options or models use the same machines and people, mixed-model-sequencing production enables a smooth flow and work content between products. Mixed-model sequencing also minimizes quality problems because fewer numbers of a particular model are built at one time. In addition, the line produces a lower-cost product with a higher-quality focus using the predictable mixed-model sequence, or a repeatable sequence coming on a regular basis. Also, imbalances in work content among options on a line are less disruptive, and sales can actually promise a mix of orders within a day. For example, if the weekly demand equals 77 units of the standard model, 25 with option A, 10 with option B, and 3 with option C, the mixed-model-sequenced plan might include 15 standard units, 5 with option A, 2 with option B, and 1 with option C.

On the second day, the plan is the same, except option C is omitted. The first- and second-day plans alternate every other day. Under such a plan, the units are not in a batch but are mixed throughout the day.

Sequencing Rules

The sequencing of sales orders and finished-goods orders into a flow line (as previously discussed) is very common in mixed-model Demand Flow manufacturing. This method of sequencing accommodates the variety of different products produced in the same flow line. A mix of models also allows a wide variety of products to be manufactured within the space available on a production floor. Productivity results from effective capacity utilization, and the products manufactured each day should be those required by customers and sales.

To maximize operations' ability to manage variety and produce TQC units to a linearity target, well-defined sequencing rules must provide flexibility and acknowledge the constraints within the production environment. Each production line has its own constraints, requiring different sequencing rules for individual flow lines. There are only four main objectives for these sequencing rules.

The first objective of sequencing rules is to provide balance within a flow line by sequencing products with different actual sequence-of-events times. World-class manufacturers have found that many different sequencing rules apply to different product types and mixes, but overcoming imbalances is critical to the linearity of production. Sequencing products in a machine environment to aid productivity with different setup requirements is a common approach.

The sequencing rules provide production planning with the desired mix and sequence to minimize required setup time given a particular mix. The sequencing rules associated with the last two issues—batches and customer-quoted lead time (CQLT)—typically apply to process industry environments.

In this situation, sales sells products in quantities that are lower than the production-batch quantity, but sales also desires to minimize finished-goods inventory created by producing a batch of product based on a small sales quantity. The sequencing rules allow production planning to hold the sales orders and look for a buildup of orders for the same product that approach that batch quantity. If the batch quantity is sold across multiple orders, then the production begins immediately and the sales orders are satisfied via shipment. If the CQLT is in danger of being violated before the cumulative sales-order quantities have reached the batch size, then production of the batch is initiated to meet the CQLT and finished-goods inventory result. The sequencing provides a stable set of rules for a consistent approach to the release of work to the production floor, resulting in the best mix from an input of sales-order demand. Notice that no "standard" sequencing rules exist; instead, sequencing follows the four main objectives that the production-planning team uses to best support the production operations.

Why Linearity?

In traditional scheduled manufacturing, the primary objective is to meet the monthly goal, regardless of when it is met during the month. Some traditional production charts resemble hockey sticks, with little production through most of the month until the last few days when virtually all production takes place.

In other charts, trend lines sag, starting and ending near the average daily rate but dropping off substantially for several days in between.

Demand Flow manufacturing emphasizes linearity for several reasons. Wide variations in the daily rate usually require overtime to catch up. That overtime is a combination of additional overhead and labor costs of 1.5 times the normal rate. A wide variation in daily production generally causes new quality problems or aggravates existing ones. Finished-goods analyses consistently reveal that units produced at the end of a hectic production month have nearly twice the defect rate as units produced at other times.

Consistent production tends to promote consistent quality. Consistency also raises customer satisfaction and service levels. Sales can promise an order to a customer on a specific day and reasonably expect the unit on that date, rather than a vague promise of sometime later in the month. Also, the short-term focus of a daily rate simplifies performance monitoring and makes goals more attainable. The shorter the term of the goal, the more likely it is to be met. Linearity is substantially more cost efficient. It produces higher quality, higher customer satisfaction and service, and more realistic production goals.

Measuring Linearity

Performance against the master production plan should be measured on a daily basis. Actual production compared against the planned daily-production rate yields deviations. Deviations should be tracked on the basis of the following: deviation for the day, sum of the absolute deviations for the month to date, net deviations, and the cause of deviations. Offsetting deviations should provide no comfort, even when at the end of the month the net deviations column stands at zero.

For example, consider a nine-day period during which the production rate is four per day and the actual production volumes are as follows: 3, 4, 5, 3, 4, 5, 3, 4, 5. The production rate and actual rate both total 36. Traditionally, the net deviation is zero, because the overproduction and underproduction cancel out each other. However, the sum of the absolute deviations is 6. The sum of the absolute deviations—a cumulative figure that can only increase—serves as the basis for calculating the linearity index.

In the example above, the linearity index formula is: 1 minus [6 divided by 36] times 100 percent, for a linearity index of 83.3 percent. Flow manufacturers should not seek daily delivery of supplies until a linearity index of at least 90 percent is achieved. Each cause of deviation should be listed and analyzed. The objective is to eliminate such causes and achieve greater linearity.

In summary, the goal of the world-class manufacturer is to be a demand-based producer in a flexible Demand Flow process. Total-quality products are built in a linear flow equal to the targeted daily rate. It is a major accomplishment to achieve adherence to customer demand. Flexible time-fence and planning-flex-fence controls are essential to allow sales to make the necessary small changes in a short period of time and eliminate the large spikes prevalent in traditional scheduled manufacturing. The likely results: high customer satisfaction and service levels, with a minimum of inventory investment and a flexible, cost-effective Demand Flow Technology production process.

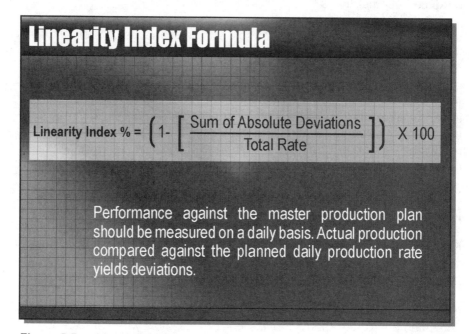

Figure 5.5 — Linearity Index Formula

Chapter 6
Kanban Pull and Backflush Techniques

In Western countries, the philosophy that pulls material on demand is called "just-in-time inventory." In Japan, it is often referred to as "kanban manufacturing." Kanban is a very powerful technique that is more than simply a tool to execute the pull philosophy. It is also more than its Japanese-to-English translation of "communication signal" or "card." Kanban is the primary tow chain in a demand-pull system. It replaces the reams of paper associated with scheduling and issue actions. It is frequently the key to saving many millions of dollars per year because of the substantial reduction in time required to carry materials in inventory, as compared to traditional schedulized manufacturing.

Demand Flow Technology moves away from the philosophy of scheduled manufacturing and toward the kanban demand-pull philosophy—where the material is pulled to the point needed, when needed, and no sooner. Many companies have adopted kanban techniques and methods, some with limited success. This limited success is surprising, given that the basic kanban methods are designed for easy deployment and use. The elements of a kanban material-pull environment described below enable an understanding of the basic tools and how they can be employed.

Raw In-Process Inventory

In Demand Flow manufacturing, points of replenishment (resupply), inventory supply areas, and carts or stations sit close to the consuming line or feeder processes they serve. It should be emphasized that these supply areas stock material at or near where it is consumed. When the consuming line needs components or materials, it pulls from these points of replenishment supply areas.

These supply areas in turn pull from a storeroom (while one still remains at the plant); but ultimately, a pull directly from an external supplier replenishes these supply areas. No computer transaction ever takes place between the point-of-replenishment supply area and the consuming line. In other words, parts should never be counted between those points. From the standpoint of inventory management and cost accounting, the line and the point-of-replenishment supply areas are all in "raw and in process" (RIP) inventory. All of the materials and parts, whether they are on the line or at various supply points, are considered to be in RIP. Materials kept in a storeroom, in either traditional or Demand Flow manufacturing, are still classified as raw material. The movement inventory from a storeroom (RAW) into RIP requires a physical count and a computer account transfer (see Figure 6.1).

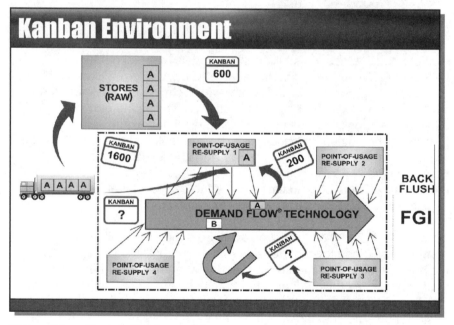

Figure 6.1 — Kanban Environment

RIP Inclusive

For purposes of inventory, RIP excludes all RAW materials and considers anything from completion of the product back to the point where supplies first enter the process as RIP. Material is pulled into the process, either from a storeroom or directly from a supplier. What comes into RIP from suppliers or the storeroom is the subject of a computer transaction. A receipt transaction acknowledges the specific quantity of materials from the supplier into RIP, or from the supplier into the storeroom. However, the movement of material from any RIP point to any other point in the RIP environment is not tracked. Everything in the point-of-replenishment supply areas and on the line is considered RIP inventory.

At this point, RIP has approximately five days of raw parts, as a rule of thumb. This rule begins the very first day of flow-process design and initiation. In Demand Flow manufacturing, the exact quantity of material within that RIP area is known, but since the amount is so small, the exact quantity at each location is not a topic of concern. What *is* known is the exact total inventory count in RIP in aggregate at any given time.

In Demand Flow manufacturing, no counting or computer transaction takes place between the point-of-replenishment supply areas and the consuming flow line.

Backflushing for Relieving RIP Inventory

Once the completed product passes from production to finished-goods inventory to become available for shipment, the product bill of material is backflushed to relieve the component parts from the RIP inventory. This backflushing technique is a computer and financial transaction. As completed quantity of finished products comes to the back of the flow line, a computer completion transaction is processed for the quantity of the product. The computer uses the bill of material for the product and "explodes" the quantities of all the component parts in the bill of material, using the quantity of finished product multiplied by the quantity of the component required to make one of those products. The result provides the components and quantities consumed in production, assuming the bill of material is accurate. The computer then uses these component quantities to relieve (reduce) the inventories of the components in RIP inventory.

The computer also updates the finished-goods inventory with a receipt transaction for the finished-product quantity and creates the associated costing transactions. Backflushing is the means by which product completion and material consumption are measured, with a minimum of counting and paperwork. A backflush transaction must occur when completed products go into finished-goods inventory. Once a standard bicycle, for example, reaches finished goods, backflushing automatically takes from inventory the bicycle's bill of material (i.e., two tires, one handlebar, one seat, so many spokes, and so forth).

Since inventory in RIP is considered as one large bucket with approximately a week's worth of inventory parts and dollars, it is not practical to actively cycle count in the flow process. Quantities in RIP are auditable, but they should not be cycle counted on a routine basis. If accounting feels the need to audit inventory in RIP, they can audit the material in RIP.

This includes all material in the process prior to backflush, whether it is in the point-of-replenishment supply area, in a line kanban, or contained within a partially completed product on the line. Although a flow production process can be audited, it should not be actively cycle counted as in traditional scheduled manufacturing.

Eventual Elimination of Stores

Although an important goal of Demand Flow manufacturing is to eliminate the central storeroom, this does not mean taking it apart and scattering the remaining inventory throughout the production area. Eliminating the storeroom can only happen once the combined RAW and RIP inventory turns have reached 50 per year. Inventory turnover calculations refer to the total annual inventory investment for a part, divided by the average total amount of inventory for the part in all areas, including the following:

- Line
- RIP
- Finished goods
- Stores

If the total annual material cost for a particular product is $1 million, and if on-hand inventory of parts to produce the product equals $100,000, then the inventory turnover for that product is ten per year.

With an inventory turnover of ten, a storeroom needs to store material. Approximately a week's worth of purchased materials can be held in RIP.

More than a week's worth of most material requires a storeroom. In the example above, one/50th, or approximately $20,000, of that material is kept in RIP. The additional $80,000 is kept in the storeroom. If $20,000 becomes sufficient for the total inventory on hand, the parts no longer require a storeroom, and the material is pulled directly from the supplier into the point-of-replenishment supply area, rather than from the storeroom. At that point the number of parts used for the product has reduced the need for the storeroom.

Quality problems, delivery problems, and the lack of a DFT contract with suppliers are additional reasons for maintaining a storeroom. Since the Demand Flow process does not produce subassemblies or pick kits of parts (as in traditional manufacturing), the central storeroom function of picking parts based on a predetermined schedule is eliminated. RIP contains only approximately a week's supply of parts, and the balance of parts remains in the storeroom. RIP requires a degree of trust in employee management of the flow process.

Meeting that objective not only stipulates that employees understand RIP requirements, but also that they are capable of managing these requirements and monitoring the results.

Extremely Hazardous

Some practitioners have not only eliminated their storerooms, but they've also failed to use point-of-replenishment supply areas. They take supplies directly from the trucks to the consuming lines, which can be very dangerous. The direct supplier-to-line connection requires substantial trust, a total-quality part, prepackaged parts, 100 percent on-time delivery, 100 percent supplier certification, and a flow production process of 100 percent yield. Although meeting these criteria is possible, it is initially preferable to deliver to the point of replenishment in RIP, and to pull from there directly to the consuming line. Eventually—with a TQC part, TQC supplier, and TQC process—the supplier can deliver the part directly to the consuming line.

Kanban Techniques

Kanban is a technique that signals replenishment. Though basically a uniform card, the kanban can assume several different shapes and functions. The kanban signal can be a card, container-based electronic communication, or a combination of multiple cards.

When the kanban card is attached to a container, the container is referred to as a "kanban container." The basic kanban is a card that shows point of usage (where the part is consumed), point of supply (the point from which the part is replenished), quantity, and part number and description.

The complete information on the kanban card makes it an effective replenishment tool because it specifically describes where material comes from and where it is consumed. Except for the affected pull quantity, kanban information is very specific. As shown in Figure 6.2, the kanban card displays the following information: "Part Number 11040701; Description: Resistor, 32 ohms, 1/4 watt, 5 percent; Usage: Cell 4-3; Supply: RIP P; Pull Quantity: fill to line."

The card simply states that Cell 4-3 is consuming that particular resistor, and the resistor is replenished from a location known as RIP P. The quantity of parts pulled when the container is emptied is based on an approximate quantity in a particular container filled to a line marked inside the container. This quantity is called a "quick count" and is used on line kanban containers to simplify work for material-handling personnel.

Since no computer transaction occurs following the replenishment of a line kanban from a RIP location, it makes no sense to ask for an accurate parts count during the replenishment process.

Kanban containers on the line are selected to suit the component part. Container size accommodates the calculated kanban size. This often results in quick counts such as "fill to line," "fill container," "two scoops," and "handful," to indicate the amount required to replenish the line kanban. Of course, the kanban quantity in the container can be approximate since the quantity represents the replenishment-time interval between the line and RIP.

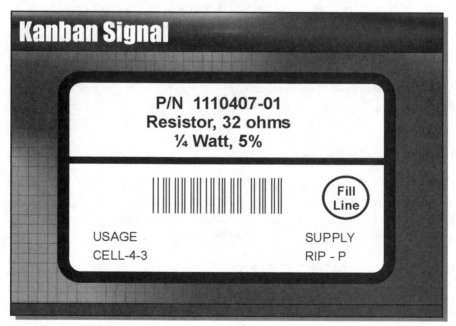

Figure 6.2 — Kanban Signal

Signal for Replenishment

Kanban is a technique of the Demand Flow manufacturing material-replenishment system. Using a single-card/container kanban technique, a kanban container must be refilled as soon as it is emptied. The container is replenished from the point of supply listed on the kanban card. The quantity indicated is pulled back from the point of supply to the point of consumption listed on the kanban. The kanban pull process is defined as the points of supply and usage of a particular part. Thus, all of the information required to replenish the kanban is listed on the card. A material-handling employee usually "roams the production process" and replenishes raw-material kanban containers. The average amount of time required to make a "milk run" of the process and replenish kanban is known as "replenishment time." This time is a key ingredient in kanban sizing.

Kanban is not tied to a specific product kit or work order. Instead, it ties to specific points of usage and supply. Multiple products built on the same line use the same component part number that was pulled from a common kanban container.Kanban cards should not be identified with the specific product that uses the particular component part.

Kanban Pull Sequences

Many different kanban pull sequences can coexist, with each expressed in terms of point of use and supply. In addition to kanban between points on the line and the point-of-replenishment supply area, kanban signals are also used to pull material between RIP and storeroom, between RIP and supplier, and between supplier and storeroom.

The kanban signals travel back and forth as needed from point of usage to point of supply. The kanban returns with the identified materials pulled and ready for consumption.

In-Process Kanban (IPK)

Kanban also can correct an imbalance between two consecutive operations in the flow process. The in-process kanban does not have a part number identity. It can be as simple as a square with a large "X" on it. An in-process kanban between operators alleviates an imbalance in the line. Thus, the common material kanban (used to pull material) differs from an in-process kanban (a buildup of units to allow an imbalanced operation to work additional hours of overtime or an additional work shift).

Kanban Pull Quantities

Counting and verifying a quantity of parts moving between the supplier and storeroom, between the supplier and RIP, or between the storeroom and RIP is essential. However, since no inventory transactions occur within the RIP area, there is no need to count specific quantities of parts within that area. Furthermore, since counting parts adds no value, there is every reason to avoid counting parts on the line. Pulls from the point-of-replenishment supply areas in RIP to the line should be as quick and efficient as possible. Thus, the preferred "quantity" shown on the line kanban would result from a quick count quantity, as discussed previously. Minimum/maximum quantities should never appear on a kanban and should also be avoided. The minimum/maximum techniques of traditional manufacturing have no place in an efficient kanban pull system. They require more detailed measurement, and they continually require counting and recounting.

The kanban pulled from a supplier includes a specific pull quantity. It is acceptable to pull the exact kanban quantity or less—never more. Depending on part usage, physical size, supplier quality, delivery, and other factors, the quantity pulled to the line with a kanban may represent several days, one day, or just a few hours of usage.

Where possible, reusable containers should be used as the kanban itself and serve several additional purposes. Such containers, filled as directed by the attached kanban card, simplify quantity determinations, reduce supplier-packaging costs, and reduce packaging waste in the plant, in addition to serving as the kanban.

Single Card Kanban/Container Techniques

With a single card/container technique, the consumption of a container of parts triggers a replenishment of the kanban quantity for that part. When the production operators empty the first container of parts, they place the empty container in a properly identified area nearby.

They immediately pull a second full container of the same part, located at the line, normally behind the first container—and continue to build the product. Thus, the most popular kanban method is the single card/ container technique, with two containers and identical kanban labels attached to each container. These kanban containers rotate between the line and RIP as material is consumed and requires replenishment. The material handlers that "roam" the RIP areas replenishing kanban take the empty container and read the attached kanban information. They then go to the identified point of replenishment (supply area) and fill the container with the quantity of parts identified on the kanban. They return the filled container back to the consumption point and place it behind the container currently being used by the production operator. If each container holds one day's worth of parts, the material handlers have that amount of time (one day) to replenish the container. If each container has two days' worth of parts in it, the material handlers have two days to replenish the container, and so on.

This is an example of the single card/container technique using two containers. Additional containers (two, three, four, or more) could be used when employing the single card/container technique. When each is consumed, it is replenished.

The only time a single card/container technique, which uses only one container, can be effectively used is when the production operator is responsible for replenishing the material. In such cases, the material-replenishment time should be added to the TQC sequence-of-events time to produce the product.

Dual-Card and Multiple Card Techniques

Using a dual-card kanban technique when containers of parts are pulled may not trigger an immediate replenishment of that part. The dual-card technique uses a "move" kanban and a "produce" kanban card. The move card pulls a quantity of material from a point of supply back to the consuming process. The produce card identifies the replenishment quantity of parts produced to satisfy the consuming demand. Material is being pulled in one (move) quantity and being replenished in another (produce) quantity.

The dual-card technique is most common in machine cells with lengthy process times or with yield or setup issues that prevent a machine from producing in much smaller move quantities.

For example, consider a machine cell that produces several different parts for various consuming flow lines (see Figure 6.3). Because of a lengthy heat-treat operation, this cell cannot produce parts in quantities of fewer than 75 pieces. Parts manufactured in this cell are taken to a point-of-replenishment supply area, RIPC1. Several different flow lines pull a part number—designated as part "A"—that is produced in the machine cell from the same supply area, RIPC1.

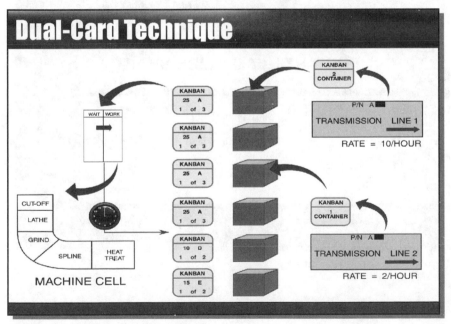

Figure 6.3 — Dual-Card Technique

When part A is manufactured, three containers of 25 each are produced and taken to RIPC1. Each of the containers has a "produce kanban" attached. Each produce-kanban card has a quantity of 25, with a "1 of 3" notation on each produce-kanban card (see Figure 6.4).

When the flow lines consume the parts, the material handlers read the move-kanban card and go to RIPC1 to pull a container of parts back to the consuming line. When the part is pulled from the RIPC1 supply area, the handlers remove the produce-kanban card from the container of parts and place the produce card in a collection box. Throughout the day, the produce cards are removed from the collection box and taken back to the machine cell. When the machine cell gets three produce cards, production begins on another quantity of 75 of part A. The parts are being moved in containers of 25 and replenished in quantities of 75. It is very likely that at least four kanban produce cards all have the same "1 of 3" notation on them. This allows a container of 25 to remain in process while additional parts are being manufactured in the machine cell.

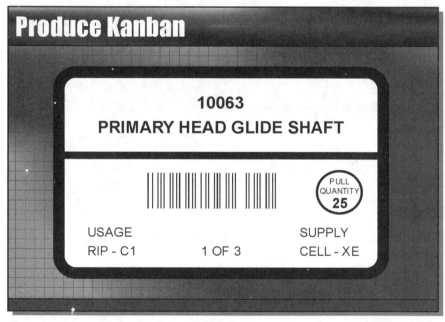

Figure 6.4 — Produce Kanban

Kanban Sizing

Kanban size is first calculated when the process is designed. The process design uses demand at capacity, which is the highest required rate. Although in some situations the kanban sizing calculation can also use this rate, demand at capacity is usually not the right volume to use. Demand at capacity is often defined as higher than the current sales volume.

This is a prudent selection when designing the flow line, with companies planning to build in extra capacity to accommodate growth over the next year or so. If the subject line design includes products expected to grow in demand, then the demand at capacity can often represent a 20 percent increase over current volumes.

This causes kanban demand at capacity to correspond to the maximum current demand rate to support sales volumes, with the expectation that the kanban sizing may need to be repeated as sales volume changes. Normally this recalculation is only necessary once each year, with the kanban demand at capacity selected accordingly. The quantity developed through the kanban sizing formula is the minimum amount of material in the process to support this rate. Fewer than the minimum quantities result in shortages in the production process before replenishment of the kanban.

The formula is used for all line, RIP, and supplier-to-stores kanban containers. Once the size of kanban has been established, it is preferable to make adjustments by increasing or reducing replenishment time. Without a permanent and dramatic change in demand, it is preferable to avoid frequent kanban resizings, as discussed above. To calculate the kanban size, divide the result of the total demand of each product per shift (D) multiplied by the usage quantity per product bill of material (Q) multiplied by replenishment time (R). To express replenishment time in hours, divide R by the number of hours in a shift available for replenishment activities (H). To use a prepackaged container quantity for a kanban quantity, divide that demand quantity by work hours per shift (H) multiplied by supplier package quantity (P). This supplier package quantity is optional:

$$\text{Kanban Size} = \frac{\Sigma (D \times Q) R}{H \times P}$$

D = Kanban Demand at Capacity Per Shift

Q = Usage Quantity Per Product

R = Replenishment Time (hours)

H = Replenishment Work Hours Per Shift

P = Supplier Package Quantity (if applicable)

Consider a part with the following characteristics: kanban demand, 120 units; usage quantity per unit, 17; replenishment time, 2 hours; replenishment work hours per shift, 7.5; supplier packaging, 25 per package. The minimum kanban quantity is 120 times 17, times 2, divided by 7.5, times 25—or 22 packages. That results in 550 pieces. If there is yield loss on a component, that loss should be factored into the usage quantity.

The total kanban demand of a part should be used for kanban sizing. For example, consider three products built on a mixed-model flow line. Product X has a kanban demand of 50 per day. Product Y has a kanban demand of 40 per day. Product Z has a kanban demand of 5 per day. Each product uses one component part W at the same consuming location on the line.

The minimum size of the line kanban in part W receives a replenishment time of four hours and is sized as follows:

$$50 \times 1 = 50$$
$$40 \times 1 = 40$$
$$\underline{5 \times 1 = 5}$$
$$95$$

Total Demand = $\dfrac{95 \times 4}{7.5}$ = 51 pieces

Thus the exact kanban calculation is 51 pieces. With approximately 60 part Ws per measured cup, the line kanban contains a pull quantity of one cup. Unless a special need requires a precise count of part W, the quick count quantity is used between the line and the RIP supply point.

When a new product is added to the process without significantly increasing the total rate per shift or the total usage quantity of the material already in the process, there is no need to add quantities to existing kanban containers or change the size of these kanban containers. However, if new material requirements or a significant change in the total daily rate per shift or usage quantity occurs, existing kanban sizes may be increased, or multiple kanban containers may be added.

Non-Replenishable Kanban

Some situations require a different type of kanban. Some kanbans pull specific component materials or parts to the consuming flow line only when the parts are required. Most often, this kanban technique applies to unique component parts required in low-volume products. Examples of these parts include specific build-to-order or procure-to-order components, where an inventory of these parts is inappropriate for RIP inventory on the production floor.

Because these parts are low volume, they are pulled to the line only when required to build a specific low-volume product. In some cases, the company has procured these unique parts to suit a sales-order specification.

In this situation the production-planning team identifies the parts required from the sales-order configuration and communicates to the storeroom the requirement to pull the specific part to the flow line. The communication used is the non-replenishable kanban.

In most cases, this is a card that identifies the part number, description, point of supply (storeroom), and point of consumption (line). The actual quantity of the part number often is written on the card by production planning to suit the sales-order quantity being manufactured. The kanban card is released by planning and authorizes a material handler to bring the material to the flow line. When the product is produced and the non-replenishable kanban part consumed, the kanban card can be returned to production planning or destroyed. If the production-planning team identifies that the part is very rarely used, they can designate that the non-replenishable kanban card be printed anew each time. If the requirement will likely reoccur, albeit infrequently, the team may choose to have kanban cards returned to be used again.

Kanban Manager/Planner

With the introduction of kanban techniques and methods within a manufacturing environment comes the necessity of appointing a kanban manager or planner, usually within the materials-management team. This individual's responsibilities encompass the management of the entire kanban environment, providing a focal point for all kanban control within the facility.

This appointment is critical to the success of the implementation of kanban and to ensure excellent understanding and maintenance of materials control.

As the kanban and DFT environment matures and the company achieves linearity performance, the expansion of the kanban environment upstream to the suppliers becomes a priority, and DFT supplier contracts and demand negotiations commence to provide extended benefits to the manufacturer and the suppliers.

Backflushing the Bill of Material

The Demand Flow manufacturing bill of material also assists in inventory control on parts in RIP. A computer technique referred to as backflush relieves the component on-hand inventory from RIP of the component parts used to build a product. The computer backflush transaction takes place when the final product is completed. The product bill of material is deducted from the on-hand RIP inventory by the computer backflush transaction. The basic assumption of the backflush technique is that if the product was built, the bill of material for the product must have been used.

Backflush Information

A backflush transaction requires some additional information on the DFT bill of material beyond what is shown on the traditional product bill of material. The DFT bill of material for a product includes not only the component part numbers and quantity per unit but also identifies the backflush location and deducts identification. The backflush location identifies whence the component part is to be deducted when the product bill of material is backflushed. The backflush location for products produced in a single facility is usually identified as "RIP." The component part number and its associated "quantity per" are relieved from the on-hand quantity at the backflush location shown (see Figure 6.5). It is possible that inventory from the DFT bill of material could be consumed at different backflush locations. If such instances occur, the product bill of material shows a component part number listed twice, with different quantities and different backflush locations. This scenario would apply if a manufacturer had two different production buildings with some parts consumed and backflushed from Building 1 RIP and others consumed and backflushed from Building 2 RIP.

Products built on multiple lines may have separate and distinct backflush locations—but always only one product bill of material. In that case, a line identification number tied to a product defines the backflush location for the product being produced. So when the computer completes the back-flush transaction, it specifies the product part number and the line identification. The computer system obtains the appropriate backflush location based on that information.

Bill of Materials (BOM) Planned Usage

The underlying principle of backflush is planned usage. In other words, if the product was built, the product bill of material was consumed. This requires a high degree of accuracy in the bill of material, as well as the requirement that any part that is controlled in inventory must be on the bill of material. Thus, packaging and other traditional expense items should be put on the bill of material to the extent possible. If component usage can be consistently predicted, then it should be on the bill of material as well.

These practices should eliminate traditional techniques, such as order point and minimum/maximum. The bill of material assumes primary control when usage can be predicted. The DFT bill of material must also be accurate and complete.

Backflush at a Deduct Point

The purpose of an intermediate computer backflush transaction is to take a portion of the product bill of material and relieve only those parts from the RIP inventory that have been consumed prior to completion of the final product. An intermediate backflush takes place at a physical location in the production process, known as a "deduct point."

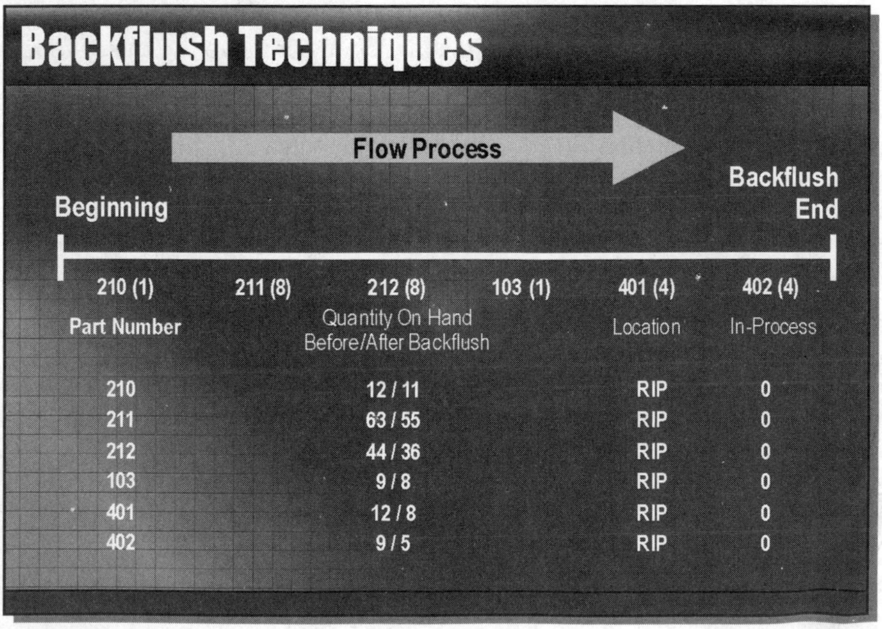

Backflush Techniques

Flow Process

Beginning / Backflush End

Part Number	Quantity On Hand Before/After Backflush	Location	In-Process
210	12 / 11	RIP	0
211	63 / 55	RIP	0
212	44 / 36	RIP	0
103	9 / 8	RIP	0
401	12 / 8	RIP	0
402	9 / 5	RIP	0

210 (1) 211 (8) 212 (8) 103 (1) 401 (4) 402 (4)

Figure 6.5 — Backflush Techniques

With an intermediate backflush, the parts relieved from the RIP inventory are placed in a special holding bucket called "in-process backflush inventory." When a component part has been placed in the in-process backflush inventory, its bill-of-material quantity has been relieved from RIP— but these component parts have not yet reached the end of the process as a part of the final, completed product.

At the time of an intermediate backflush, the parts consumed up to the physical deduct point are relieved from the backflush location of the flow-manufacturing bill of material and placed into an in-process backflush inventory.

Final Backflush Follows Intermediate

When the product is completed and placed into finished goods, a final backflush is done against the entire product bill of material. The portion of the product in the in-process backflush inventory is taken from that in-process inventory "bucket," and the portion of the components on the bill of material not yet relieved at the intermediate deduct point are taken from the backflush location.

The rule for this computer transaction is simple: if the component-part quantity is in the in-process inventory bucket, take it from this in-process inventory; if it is not on hand in in-process, relieve the inventory from the appropriate backflush locations. This way, the total on-hand inventory balance from a formal computer system standpoint can never be misstated.

It cannot be assumed that inventory consumed at the intermediate deduct point is no longer on hand in RIP. This assumption causes the purchasing planning system to assume that those component parts were used and to create an immediate replenishment for those component parts. From the standpoint of the formal computer system, those parts are still considered on-hand inventory. They are simply in an in-process holding inventory bucket waiting for the final backflush of the completed product.

Not a Mandatory Transaction

It is not essential to have any intermediate backflush or deduct point in the flow production process. However, such points may be desirable. Sometimes, in very lengthy processes where the total product cycle time exceeds three days, the intermediate backflush may be necessary to assist in managing inventory control. Under these and certain other circumstances, an intermediate backflush or deduct point can be quite advantageous.

For example, if 95 percent of the parts are consumed prior to a lengthy burn in operation (perhaps ten days), an intermediate backflush of the 95 percent does two things:

1. It allows an audit of the component parts up to the burn in operation, without having to count parts in the burn in operation.

2. It allows any non-repairable failures from the burn in process to be backflushed to scrap as a single entity rather than as several independent parts.

In this example, the remaining five percent of component parts consumed after the burn in operation are backflushed as usual at the finished-goods stage. Intermediate deduct points should not be used without good reason. They are not recommended unless there is a scenario similar to the example outlined. While intermediate deduct or intermediate backflush points are not necessary, lengthy processes tend to have at least one of these points for effective inventory control and auditing purposes.

Scrap

Scrap is one area that Demand Flow manufacturing treats very differently than traditional scheduled manufacturing. Visual techniques are used, and the material handlers process the paperwork. Operators are free of the scrap paperwork morass. A preferred method of keeping the operators free of paperwork involves the use of a poker chip.

When operators find a bad part, they place a poker chip in their line kanban. Each poker chip in the kanban represents one piece of defective material. When material handlers replenish that empty kanban, they note the number of poker chips and write the appropriate scrap tickets. Because the backflush transaction deducts only the planned consumption of a component part, any unplanned consumption, such as extra usage or scrap, must be reported.

The scrap transactions are reported against parts. In the case of a non-repairable final product, all of the consumed component parts can be backflushed from RIP inventory into a scrap account for the final product. In Demand Flow manufacturing, scrap must be handled separately. It is not part of the planned backflush.

Scrap must be processed outside the backflush transaction of the DFT bill of material. Predicting yield on a particular component part requires extreme caution. If the yield is above or below the prediction, then the on-hand inventory is misstated. Failure to report scrap in a timely fashion makes it impossible to maintain correct inventory balances.

In scrap transactions, only the material value of the component or the scrapped assembly is applied. No partial labor or overhead credit should result from a partially assembled unit or from partial completion of the production process. Transactions should be made only against the material portion of the component or assembly.

Communication to Pull

Kanban is the method of moving material into the flow-production process, as compared to traditional manufacturing's work-order issues. kanban is a communication technique that specifies where to find material, where to take it, and when to pull material. Kanban sizing is calculated at the minimum kanban quantity necessary to support the designed product or mixed-model line rate. Although the volume of products produced can continually change, the kanban quantities are rarely affected or resized. Single- and dual-card kanban techniques are vital tools to the competitive world-class corporation.

Chapter 7
Group Technology and Machine Cells

Group technology is a technique in which dissimilar machines are grouped into single cells to produce families of similar products. The products produced in these cells use various machines or operations, but each product may require slightly more or less work time per machine or operation within the cell. Group technology organizes people and machines into cells to produce lower quantities of parts with reduced setups, reduced or eliminated queues, and reduced throughput times. Group technology also provides the opportunity to use flexible operators for multiple machines, facilitating such efficiencies as an operator performing setup work on one machine while another machine is still running. For example, one to five operators can run a machine cell of five different machines, depending on the machine run time.

Machine Cells Provide Flexibility

In the most basic configuration of scheduled manufacturing, machines are grouped into machine work centers according to function. Each of the similar individual machines within the functional work center is called a workstation. Moving from functional work center to functional work center creates products. For example, a part that requires machine stamping, grinding, and drilling starts off in the first work center performing the first operation, which is stamping.

All parts are stamped, and when the entire lot is stamped, that lot is sent over to the grinding work center. All parts are then ground in the grinding work center and then transferred into the third work center for drilling. Some scheduled manufacturers have adopted the group-technology approach of creating cells of dissimilar machines as described above, but designate a work-center identity to the entire cell.

These manufacturers still schedule batches of parts through these work centers and often produce batches that are just as large as they produced prior to the creation of the cell, which only results in maintaining the status quo.

In Demand Flow manufacturing, machine cells consist of dissimilar machines producing a family of products that require the combination of various machine functions, such as stamping, grinding, and drilling. The flow machine cell might include one stamping machine, one grinder, and one drill rather than grouping all machines of one function, as is the case in the traditional work center. It is common to employ a U-shaped machine arrangement in designing a machine cell. The U shape allows for the operation of a maximum number of machines in a relatively small area and enables one production employee to move easily from machine to machine with a minimal amount of move time within the cell (see Figure 7.1).

When a product leaves a flow machine cell, it is complete in terms of the operations performed on it. In the example above, when the product leaves the machine cell, it is stamped, ground, and drilled. To achieve the same result in traditional manufacturing, the product is scheduled, queued, and moved or routed to and through three completely different functional work centers.

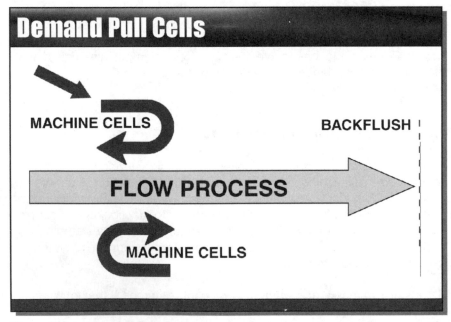

Figure 7.1 — Demand Pull Cells

Staffed by Flexible People

At a minimum, the multifunctional employees who work within the machine cell should be capable of moving up and down through the various machines within the machine cell. A machine cell can consist of a single person or several people. Machine cells that are feeder cells should be arranged at or near their consuming production line. As machined parts are pulled into the consuming line, they are immediately verified and consumed. This improves communication and quality control, and all parts produced are verified one at a time, not by the scheduled and inspected batch quantity. The multifunctional employees move similar products smoothly through the cell. The group technology techniques to create machine cells, in conjunction with the balancing techniques of the flow process, offer many options in line designs.

Cells are set up as close to the consuming line as possible, and if a U-shaped layout is used, the cell ends near where it begins as a result of that design and layout.

Demand Flow manufacturing machine cells tend to be considerably denser than the traditional work centers, and facility requirements—such as air conditioning, power, air pressure, lighting, waste and environmental requirements—are just a few of the considerations in designing and locating a machine cell.

Mixed-Model Design

In achieving an efficient cell design for multiple product cells, process-mapping techniques help ascertain the commonalties of process and product. The first step in setting up a mixed-model machine cell, or flow line, is to organize cells according to the various machines or operations required to produce a family of products. The products manufactured within a mixed-model machine cell or flow line should be identical at the process level. The first step in identifying process commonality is creating a product synchronization for each candidate product. This product synchronization specifies the vital relationship of manufacturing processes, organized in a flow, required to build the product. The second step is to create a process map, which contains the candidate product synchronizations (see Figure 7.2).

The process map identifies product commonality at the process level. Once the type of machines required within the cell is known, the next objective is to identify those products that require the common processes within that cell. Process maps identify multiple products down one axis and specify machine processes or operations across the other.

Each product is reviewed to identify the particular processes or machine operations required for manufacturing. The next step in cell design is to create a cell configuration made up of the common machines or operations identified in the process map. The physical cell layout requires those machines or operations for products that share the identified process commonality within a machine cell or mixed-model flow line. Once the configuration of the machines or operations within a cell has been determined, then all of the products (which are possible candidates because they require those various machine processes) are identified.

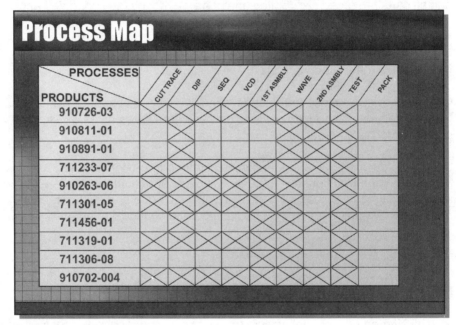

Figure 7.2 — Process Map

Searching for Commonality

After identifying possible candidates for the machine cell or mixed-model flow line, each product with commonalities at the process level has its work defined using the sequence of events tool. This specifies the work content and quality criteria for each product produced within the machine cell or flow line. This work-content definition is separately defined as machine or labor work. Once the work-content time by machine function or labor is identified, the total time is calculated for each, including all of the processes and products on the process map. This identifies the total time to produce each possible product candidate based on the common processes in the cell.

The next objective is to group products that are common in total time, and then to group those similar in the time required at each of the machines within the cell. Now another matrix evolves, one that identifies similar products that require the common processes in this cell design. This matrix identifies the work content at each machine function, also taking the labor time in the cell into consideration, along with the total work-content time of each product (see Figure 7.3).

Using an operational- and total-time index, products with common proc-esses and similar operational and total times are further isolated as final candidates for additional analysis.

Mixed-Model Commonality

CELL CONF / PRODUCTS	CUT TRACE	DIP	SEQ	VCD	1ST ASMBLY	WAVE	TEST	TOTAL
910726-03	5.2	15.2	9.1	6.2	17.4	1.7	8.5	63.3
910263-06	2.7	11.6	4.6	3.0	9.1	1.7	3.5	36.2
711301-05	5.0	11.2	5.1	3.9	7.6	1.7	3.5	38.0
711319-01	3.3	12.1	5.1	3.7	5.9	1.7	4.4	36.2
910773-11	3.9	13.1	4.8	3.1	9.2	1.7	4.0	39.8
711879-09	3.6	14.4	8.8	6.1	19.7	1.7	8.5	62.8
711332-17	7.7	11.4	10.3	7.6	18.3	1.7	9.6	66.6
711911-01	8.6	17.3	10.9	8.1	17.1	1.7	8.3	72.0
910671-07	2.6	10.2	5.8	4.1	6.6	1.7	3.5	34.5
910984-02	2.4	16.4	5.5	3.9	4.3	1.7	3.5	37.7

Figure 7.3 — Mixed-Model Commonality

Takt Time of a Cell

Targeted operational cycle time of a cell dedicated to a single part or product is established by multiplying the effective hours per shift by the number of shifts per day, and dividing that by the demand at capacity of the part or product to be run through the cell:

$$\text{Cell Operational Cycle Time} = \frac{\text{He (S)}}{\text{Dc}}$$

He = Effective Work Hours Per Shift

S = Shifts Per Day

Dc = Demand at Capacity

In calculating the targeted operational cycle time of a mixed-model cell, the demand-at-capacity volume is the sum of the individual rates of each of the similar products to be produced within the cell. This designed-cell takt time is based on the designed volume demand at capacity anticipated for each of the parts or products to be created in the cell. This total demand becomes the demand at capacity for the cell, and it is divided into the effective work hours planned for the cell. This yields a cellular takt time for all products manufactured within the cell, not an individual cycle time for any particular product manufactured within the cell. This cellular takt time is critical in sizing the number of in-process kanbans (IPK) required based on any imbalances between the actual times and operational cycle time targets of the cell. This analysis has identified those products that are excellent candidates for a mixed-model cell. It is also essential in identifying the number of machines and people required to support the volume of products to be manufactured within the cell.

Equipment Utilization

Targeted or designed operational cycle time must be calculated within a cell to ensure efficient utilization. In Demand Flow manufacturing, utilization does not drive the process, but it is a management consideration. Actual work-content time, as a weighted average, is divided by targeted operational cycle time of the cell to determine machine utilization.

If one machine in a cell is not fully utilized, one option is to have that machine feed additional cells to increase its utilization. Line layouts are adjusted to allow the underutilized machine to feed multiple cells. The cell's resources are calculated by dividing the weighted actual times of the products that require that process by the operational cycle time target.

The calculation is particularly useful in that this resource calculation applies to machines, people, operations (space), and pieces. Consider that if the formula uses the machine time from the sequence of events' actual-time definition, with times calculated to a weighted average for all the products in the cell, then dividing this average by the cell takt time (operational cycle time) yields two alternative results. Either the result provides the number of machines required, or, if only one machine is used (for example, an oven or anneal furnace), then the result is the minimum number of products (pieces) to be produced at the machine.

If the formula uses the labor time from the sequence of events' actual time definition, with times calculated to a weighted average for all the products in the cell, then dividing this average by the cell takt time (operational cycle time) yields two different and useful results. The first result provides the number of operations required on the production floor (space considerations), where the result is rounded up to the nearest whole number. But if the result is a quantity with a decimal, then this represents the number of people required to meet the demand-at-capacity volume.

Attacking Non-Value-Added Setups

Both setup steps and maintenance tasks are major considerations in the design of machine cells. The reduction or elimination of setup and move time is a primary objective. Achieving this objective involves the following:

- Defining the elements of setup
- Sequencing equipment
- Methods analysis
- Advance preparation
- Machine-support checklists

During the analysis associated with the TQC sequence of events, the time associated with each setup step is identified. Once this setup has been identified and prioritized, it is methodically attacked to reduce the time required. The first step in reducing the setup time is to observe the performance of the physical setup. The team responsible for the setup-reduction task must completely understand the work needed to complete the entire setup. After the setup has been physically observed, it is video-taped for a detailed analysis. This detailed analysis results in the sequence of events required to set up and operate the selected machine.

Each step to set up the machine is classified as either an internal or external step. Internal setup steps are those performed at the machine and with the machine stopped. External setup steps are performed externally to the machine, and they can take place while the machine is producing a part. Internal machine setup steps reduce the time a machine is available to produce parts. Reducing internal setup steps becomes the highest priority. If a setup step cannot be eliminated, it should be shifted from internal to external time, if possible (see Figure 7.4).

Tooling required to produce a part should be pulled to the appropriate machine using kanban techniques. The tool number required to produce a particular part is listed on the kanban card or container. Once the part is pulled, the kanban is returned to the producing cell. The machine operator verifies whether the tool, identified on the kanban, is currently at the cell. If the tool is not at the machine cell, the kanban is directed to the tool-and-die area to be pulled and returned to the machine cell, along with the previously mentioned kanban card.

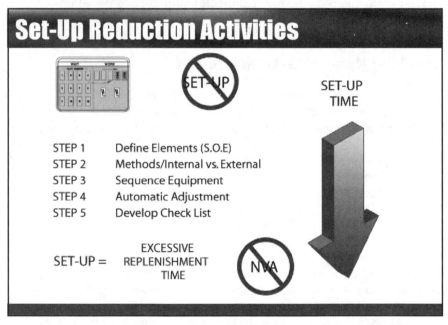

Figure 7.4 — Set-Up Reduction Activities

Checklists

Every machine should have four checklists required for its operation and maintenance. Specifically, every machine should have the following four checklists, or one checklist with four sections:

1. Setup
2. Operation
3. Troubleshooting, in the event of problems
4. Maintenance

Each checklist is specific and requires the operator to check off each required task as it is performed. This ensures the integrity of the process and verifies that the operator has completed required steps before proceeding with the machine operation. TQC operational sheets display a graphical representation of how the completed part should look, the quality criteria for the part, and the proper tools to be used. Every effort is made to eliminate guesswork and develop a consistent, predictable, and repeatable total-quality process. The first checklist shows the tasks the operator goes through to set up the particular machine. It specifically defines the appropriate sequence and the tasks to be performed. It also has a physical check box for key items that operators are required to perform during the setup tasks. The second checklist is an operational list of tasks that must be followed during the startup and operation of a particular machine.

This checklist identifies the specific sequence an operator must follow before the machine is allowed to produce the first part in the shift. The third checklist is a troubleshooting checklist. It addresses problems during the setup or the routine operation of the machine and guides operators through identifying the possible cause of a particular problem and setting verification priorities before proceeding.

Bibles of Machine Operation

If the troubleshooting checklist identifies a problem and the operator solves it, then the operator can continue with the operational checklist and bring up the machine. The fourth checklist is the maintenance checklist. This identifies the likely causes of problems that stop the operator during the setup of the machine. The maintenance checklist is designed for the maintenance worker who performs the predictive and preventive maintenance program. It identifies the likely cause of the machine's problem. Special tools and parts required to support the machine should be kept near the machine, readily available to the maintenance mechanic. The objective is to fix the machine and bring it back online as quickly as possible.

Machine Maintenance

Total preventive maintenance is a very important aspect of Demand Flow manufacturing. It is extremely important that the machines used in the process are properly maintained to ensure their availability during production. Machines should also be standardized according to type and manufacturer. Machines that eliminate internal setup and are very reliable and easy to maintain are preferable to overly large, powerful machines that perform multiple tasks.

Machine operators should perform routine maintenance (such as checking pressure settings, temperatures, and chain or sprocket tensions) before the equipment is started for the production shift. The machine operator runs through a routine checklist that identifies the specific sequence and tasks that must be verified before the machine can be operated. Once the operator has completed the routine maintenance and filled out the operational checklist, the machine can be operated.

An Integral Demand Flow Manufacturing Requirement

Total preventive maintenance is critical to flow manufacturing. A total preventive maintenance program should include the following, all of which are necessary for improved product quality and reduced in-process inventory:

- A preoperational checklist for machine operators
- Immediate verification of emergency failures against the check list—for both preventive and predictive maintenance programs
- Emergency-maintenance analysis
- Under-capacity scheduling
- Corresponding scheduled maintenance programs

The objective of predictive and preventive maintenance programs is to increase machine uptime, eliminate emergency maintenance, and make the machine more predictable over a period of time. As predictive and preventive maintenance programs improve, machine utilization improves, and in-process inventories can be reduced. Adjustments must be made for machines that routinely go down due to maintenance problems. For example, if repairing a machine and bringing it back online routinely takes 20 minutes per shift, effective work hours per shift for that particular machine should be adjusted downward by 20 minutes. This could result in a shorter designed operational cycle time for this particular machine, as opposed to the rest of the cell or flow process it supports.

This also requires a corresponding in-process kanban between the machine and its supporting processes to buffer the nonproductive time the machine is unavailable.

Demand Flow Line and Cell Designs

Demand Flow manufacturing is the technological integration of a flexible production-flow process and a demand-pull material system. The DFT production-flow process is based on in-process total quality control at the point where the work is performed. The flow line or cell layouts result from the techniques of cycle-time line designs. These flow manufacturing techniques include:

- TQC sequence of events
- Demand-at-capacity and daily-rate volumes
- Operational and total product cycle time
- TQC operations
- Kanban
- Flexible people and machines
- Linear flow-line layouts

Flexible pull lines and cells that focus on eliminating non-value-added steps are key to many of the flow manufacturing benefits: reducing work-in-process dollars, floor space, scrap/rework, finished-goods-inventory dollars, overhead, and computer activity and transactions. Product costs decrease, while product quality improves—thereby enhancing a corporation's competitiveness. Manufacturing can now become a corporation's strategic advantage!

Chapter 8
Product Design for
Demand Flow Manufacturing

Speed to market was the key strategic advantage for world-class manufacturing companies in the 1990s, and it continues to be the key strategic advantage today and into the future. The time required to develop products, manufacture them, and deliver them to customers must continue to shrink by as much as 50 percent for companies to remain competitive in a global market. Design engineering can no longer afford to be an isolated and slow-moving organization. Product designs must be cost-effectively designed for high-quality flow production and must be ready for manufacturing at the time of product release. Engineering can no longer function in a design vacuum but must assume new responsibilities as part of the total company Demand Flow manufacturing technology approach.

After introducing a product, the company must drastically simplify the release of documentation and the engineering-change process to become more effective and consistent with Demand Flow manufacturing technology.

Speed to Market

The world-class manufacturer must reduce the total time from the start of the research-and-development cycle to the completion of a released product from manufacturing. Technology turnover is the total life cycle of a product, from the time development begins to the end of the product sales cycle. The challenge to the world-class manufacturer is to minimize the development and preproduction release portion of the technology turnover cycle and to increase the sales portion. The longer a product's sales cycle, the greater the profit for the company. If a company is first to market with a product, it can charge higher initial prices and increase its market share. Eventually, when the competition enters the market, all manufacturers must become competitive producers. If the Demand Flow manufacturer can be first to market and capitalize on its quicker delivery, it can achieve a distinct competitive advantage. Once the product is released to manufacturing, it must be production-ready and functional at the time of design release. While the volume of engineering changes for new products will likely remain high, the customer's initial experience of product quality and functionality must be world class.

Technology turnover cycles are increasing as product life cycles become shorter. For example, although the United States is considered one of the technology leaders in the world, it falls well behind major competitors in the actual implementation of the technology. Many companies are unable to recoup even research-and-development expenditures through product sales because of lengthy product introductions.

Design and Manufacturing as a Team

The world-class company must break the partitions between product and process design and encourage a free flow of information, people, methods, and tools between design engineering and manufacturing. These two departments must jointly develop consistently linked products and processes to ensure the highest quality at the lowest cost with the shortest possible throughput time between manufacturing and the marketplace. Engineers must design for total quality and eliminate design inspection steps by focusing on "produce-ability." Manufacturing, engineering, and quality must collaborate early in the design cycle to ensure that manufacturing can produce a total quality product.

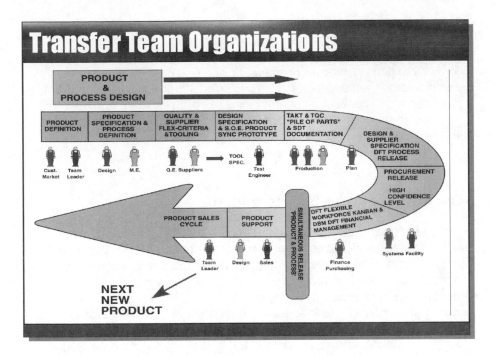

Figure 8.1 — Transfer Team Organizations

Every discipline should be deeply involved in the prototype and other cycles preceding a new-product release. Again, speed to market is the key to competitive survival.

Technology Turnover Transfer Teams

Sales initially works with outside customers to develop a product specification. The product specification encompasses the customer specification, market and competitive considerations, and price and product expectations. At this point, management selects a transfer-team leader to move a quality product to market as quickly as possible. He or she heads a cross-functional transfer team with representatives from sales and design engineering. (see Figure 8.1).

Design engineering develops a product design based on the product specifications. The product design identifies targets for cost, performance, and specifications. At this point, general component availability comes into play. Whenever possible, design engineering should use existing components selected from preferred suppliers. Based on the product design, quality engineering joins the transfer team and, along with design engineering, develops a quality specification for the product and for the process.

Quality Expectations Determined

In the next phase, the transfer team analyzes customer expectations and product specifications to verify that they are consistent. During this phase, component specifications are drafted. Customers should be invited to preview the product specifications and determine whether they meet their expectations. Supplier availability is also reviewed. If the customer response is unfavorable, the team must revisit the product-specification phase of the development process. If the product and design specifications are consistent and approved, a design prototype is produced to meet the design specifications. At this point, a manufacturing engineer and a test engineer join the transfer team. The team then begins to work on the TQC sequence of events, identifying process specifications and looking for defects that should be eliminated in the design prototype phase. The TQC sequence of events is completed during this phase.

If a total quality process flow cannot be defined, manufacturing can't produce the product and the development process must return to the product-design or product-specification phase. The product produced during the design prototype phase ensures that the product meets customer and design specifications.

Developing the Initial Demand Flow Process

If the design prototype passes this test, the next phase produces a manufacturing prototype. Talented production personnel can build this prototype in a laboratory or model-shop environment with special equipment. This prototype proves the component specifications, manufacturing documentation, process capability, and equipment tooling.

If the gifted production personnel can't produce the prototype with material bought to the component specifications—using the TQC sequence of events and manufacturing equipment and tooling—the process returns to the product-design phase. During this phase, drafting of TQC operational sheets begins.

During the manufacturing prototype phase, the capacity required to support customer demands is determined, and operational cycle time is calculated and targeted. A flat "pile of parts" bill of material is created and approved. At this point, the TQC operational sheets are reviewed, and a determination is made to ensure that the TQC design can be produced. If the test is successful, component-specification documentation can be released. Also during this phase, a planning/materials person joins the transfer team.

Design for Total Quality

The world-class flow process is one of total quality. Whenever there is more than one way to perform a task, but only one of the ways is correct, a subsequent validation step is required. Assembly tasks should not be designed to allow production employees to make a wrong decision. Symmetrical designs that allow an employee to assemble a part incorrectly require a later TQC approval in the flow process. If a person has multiple choices in performing work, but only one way is correct, and the person both completes and approves the work, the product created is classified as "designed for defect." In this example, the operation that completed the work cannot perform the TQC approval of the work.

If there is only one way to assemble a part according to a symmetrical design, then a single operation can perform and verify the work. Products must be designed using techniques that allow the total quality and verification steps to function correctly. Of course, a design that permits only one way (the right way) of performing work is preferred.

Parts Standardization is Essential

Design engineering should also have the goal of standardizing material into as few part numbers as possible. If the manufacturer already uses five different lengths of similar plastic tubing on existing products, these similar parts must be reviewed for standardization before a new-product design allows a sixth length of tubing. Design engineering should be aware of which parts manufacturing uses and which preferred suppliers are on DFT contracts. Engineering won't sacrifice product quality when they standardize with preferred suppliers. Standardization produces a higher volume per part, and typically the higher the volume of a part, the lower the price.

Focus on Suppliers

Supplier management is greatly enhanced with a smaller part-number base. As purchasing drives the total number of suppliers down and begins to focus on total cost, design engineering should focus on using these preferred suppliers in new-product designs wherever possible.

Components with long lead-time requirements are purchased; component specifications are finalized; suppliers are approved; transportation networks are set up; packaging specifications to facilitate dock-to-RIP deliveries are defined; and flexible DFT contracts are established for major suppliers. At this stage, the company has a high confidence level in the product and process design. If the confidence level is not high, a return to the manufacturing prototype phase is required.

Preparing for Release

During the preproduction phase prior to production release, all of the elements come together. TQC operational sheets are approved and released based on the targeted operational cycle time. Demand Flow lines have been designed based on the shortest total product cycle time. kanbans for both raw materials and in-process components are sized and placed. Flexible employees receive cross training one step up and one step down in the process. Any final product or process bugs are also addressed. Finance, systems, and facilities people join the transfer team to review the pre-production release from their functional perspectives.

At this time, the standard product cost is established, replacing the transfer team's previous estimates. Checklists to support the production machines are completed and released.

In the preproduction phase, a small quantity of products are manufactured and released to customers. Regular production employees build these products using the planned production machines, released tooling, components purchased to released specification drawings, and TQC operational sheets. The "real" production-flow process produces the products. Process capabilities are also validated during this phase.

The units built during the preproduction stage must meet the design specifications using components and materials purchased to the component/materials specifications. If the product meets the design specifications, minor changes can be made at this stage prior to the actual design release.

Product Release and Acceptance

The product is now ready for design release. The customer participates in the final acceptance phase of the product. Engineering change activity is high for the initial release phase, so the design members remain on the transfer team for the initial three to six months following the product release. The manufacturing flow process is now fine-tuned between operations. Employee involvement begins to improve the process and reduce in-process and material kanban closer to the calculated levels as the product and the process quickly mature. The product is designed and released for the manufacturing flow process. This ensures total quality and a product ready for production at the time of design release. The "transfer team" technology can improve a company's speed to market, offering a very powerful competitive advantage for the world-class company.

Organizational Impacts

The functional boundaries and systems inherent in the traditional product-development process are a major cause for the lengthy product introduction cycles and eventual marketplace failures. In Demand Flow manufacturing, transfer-team members report to the team leader during product development. This is not a dotted-line reporting relationship but a hard line.

The transfer-team leader must have the responsibility and authority to get a quality product to market as quickly as possible. He or she reports to design engineering until product release. The team leader, along with all team members, then reports to the manufacturing vice president or director. The design engineer and team leader work under the team leader in manufacturing for the initial period after product release. Eventually they rotate back into design engineering and onto another transfer team for the next product introduction. The primary reporting and responsibility of transfer-team members must be to the product and the transfer team, not to a functional organization.

Demand Flow Manufacturing Product Fundamentals

A flow-manufacturing environment requires that the bill of material for materials consumed in the manufacturing process be simple and easy to use. In the best possible situation, the bill of material for Demand Flow manufacturing tends to be very flat, with no subassemblies. This is in marked contrast to traditional, functional, schedulized manufacturing, which has subassemblies and fabricated parts structured into higher-level subassemblies and other fabricated parts, several times over. In traditional, schedulized manufacturing, the bill of material drives the process. In Demand Flow manufacturing, the operational cycle time drives the work content, and the bill of material becomes a basic "pile of parts." The bill of material in a Demand Flow environment is quite different from the functional bill of material now used in traditional manufacturing. The bill of material in a flow process is the fundamental building block of the product.

It is the recipe of purchased materials to be consumed by the process in order to manufacture a product. In fact, the bill of material in Demand Flow manufacturing serves three critical functions. First it defines the parts that have to be procured to produce the product. Second, the bill of material relieves the RIP inventory following production of products—the reason being that if the product is finished, the material specified on the bill of material must have been consumed. Third, the bill of material provides the cost of the material used to build the product.

Subassembly Versus Flat Pile of Parts

The world-class manufacturer strives to make the Demand Flow manufacturing bill of material approach a single level, flat pile of parts that contains only the purchased components required to build the product. Unlike traditional functional manufacturing, Demand Flow manufacturing has no subassemblies or phantoms. In essence, traditional manufacturing's multilevel subassembly competes with Demand Flow's flat pile of parts. In traditional manufacturing, function defines product design, with parts grouped into subassemblies or fabricated machine parts (see Figure 8.2).

These subassemblies and fabricated parts are scheduled, issued, tracked, and costed as separate entities. The top-level model number is the same, regardless of whether subassemblies are structured below it.

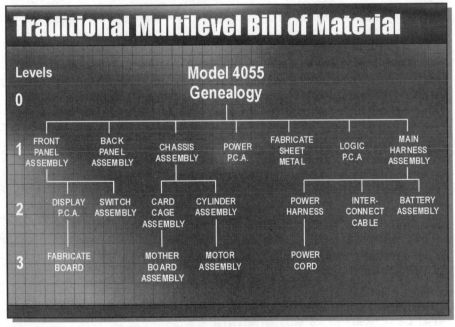

Figure 8.2 — Traditional Multilevel Bill of Material

Subassembly part numbers function simply as internal tracking and collection identities and do not add value to the product. Often, the sub-assemblies are departmentalized, with different departments building different subassemblies. In such cases, the subassemblies are sent to work-in-process storerooms and later issued to the consuming departments.

Because of this departmentalization, the bill of material drives the process layout—the design, manufacturing process, and layout are all functional. In Demand Flow manufacturing, the bill of material simply lists the purchased parts required to build a product. Where the process uses the parts is not relevant to the bill of material. Component parts or materials can be moved around, for the purposes of line balancing, without affecting the bill of material, which means process improvements can be made without complicated engineering-change orders. The result is the separation of the process and the functional bill of material.

Disconnected Parts

In Demand Flow manufacturing, the bill of material is a disconnected pile of parts, with no need for a subassembly bill of material. Operational-cycle time dictates the work content per operation, and TQC operational sheets document all operations. Demand flow has no scheduling, no picking of subassembly kits of parts, and no functional production departments. Beyond the functional design prototype phase, the multilevel bill of material has no purpose. In discussions about the "levels" of a traditional bill of material, the top—or "model"—level is referred to as level "0." The purchased parts, subassemblies, and fabricated parts directly structured to the model number are considered to be in level "1." Typically, additional lower-level purchased parts, subassemblies, and fabricated parts are structured to high-level subassemblies, creating a multilevel product bill of material.

Functional Design Versus Demand Flow Technology and Simultaneous Engineering

For years, companies have designed and manufactured products using the functional multilevel bill of material. These subassembly or machining-level part numbers are necessary to schedule, pick kits of parts, issue material, and route in-process production in traditional manufacturing. Design engineering releases these functional multilevel bills of material and maintains a documentation control system for each multilevel assembly and fabricated part. Computerized scheduling systems (MRP II) assist in the planning, scheduling, and control of these multilevel bills of material.

Traditional costing and management systems have followed suit to support this multilevel bill-of-material philosophy.

Many companies also have implemented complicated and expensive engineering change systems to attempt to control these complex multilevel bills of material.

In Demand Flow manufacturing, operational cycle time defines the targeted work content. Using kanban techniques, material is pulled to the operation where the work is performed. Since the Demand Flow manufacturer does not schedule, pick kits of parts, or build subassemblies, a multilevel bill of material has no purpose.

Because the transfer team designed and developed a product for Demand Flow manufacturing, design engineering does not require a multilevel bill of material either. World-class companies can have only one consistent philosophy for the design, development, release, and manufacture of products.

Product Bill of Material

All products that are forecast by sales or customer service and sold to customers have a bill of material, including:

- Top-level product or model numbers
- Options to standard products
- Field replaceable units (FRUs), or spare parts

Final products have model numbers that are used during the sales process. The model number is printed in sales literature, and it usually does not change during the life of a product. This model number is the product identifier for order entry and management reporting, as well as for use in communications among the manufacturer, customer, and distributors. The final product model number has a top-level part number used to control the design and manufacture of the product (see Figure 8.3).

Field Replaceable Units

A FRU (or spare part) is an interchangeable portion of a top-level product for field-service or customer-service usage. FRUs are products sold and forecasted by sales separately from the completed final product. They are not the subassemblies of traditional manufacturing. They have a separate and distinct bill of material, which may or may not share common parts with the final or original end product. FRUs are not tied to the top-level product bill of material. FRUs may be required at a later date based on the service nature of the original product. It is a totally independent portion of a final product. It can be built in the same process as the final product, or in a totally independent process. Sales sell FRUs separately, and customers buy them separately.

The product and FRU may have common parts, but the FRU is not structured as a subassembly to the final product (see Figure 8.4).

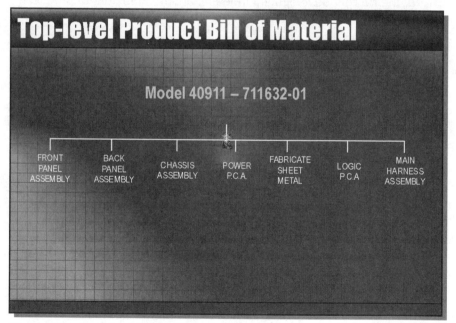

Figure 8.3 — Top-Level Product Bill of Material

An example of a FRU is a lawn-mower engine. The engine and lawn mower are originally sold together, and a similar engine is later sold separately for replacement by field service. It isn't unusual for both the product and the FRU to have some common parts, such as pistons, nuts, and bolts. The FRU always has additional costs, parts, and instruction manuals to control the replacement process by the field-service people. Additional testing and production time, shipping materials, and identification labels may be associated with the FRU, and not with the final product.

Customers may order spare printed circuit boards to repair a system installed at an earlier date. The field-service organization also may want spare components of a product, so that if a component fails, the field service person can conduct repairs at the customer's site without sending the entire unit back to the factory. Some industries and products have FRUs; others do not. Aircraft, computers, and automobiles are examples of products with FRUs.

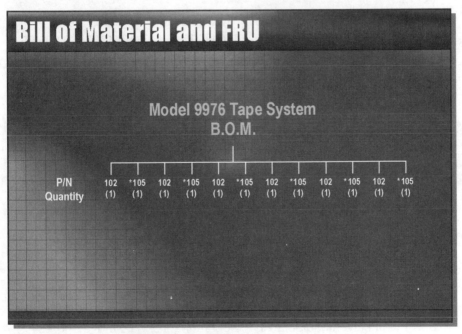

Figure 8.4 — Bill of Material and FRU

A sparkplug and wooden pencil are examples of products that don't have FRUs. The original product and the FRUs can be built on the same line or independently, whichever is preferred. FRUs can be either manufactured or purchased parts.

From a bill-of-material standpoint, the independence between the FRU and the product is further demonstrated by the fact that a change can be made at one date on a FRU and at an entirely different date, or not at all, on the final product. Additionally, the FRU is likely to contain some bill-of-material information that differs from the information on the product bill of material. That information might include a packing container, label, installation instructions, and documentation required by the FRU, which may not be a part of the original product bill of material.

Compression of the Multilevel Bill of Material

Traditional companies have almost always designed products using multilevel bills of material that are scheduled, issued, tracked, and costed accordingly. When implementing Demand Flow Technology, the world-class manufacturer hopes to flatten or compress the bill of material to a single-level pile of purchased parts. If the product has a FRU, it requires a separate bill of material. Whether or not a product has an associated FRU, all of the purchased components are structured at the bill of material of the top-level product.

One engineering change can easily compress the bill of material, thereby making subassembly drawings obsolete and bringing all purchased parts to the top-level product. Any drawing IDs of a former subassembly can be saved for historical tracking purposes. The compression of the multilevel bill of material should not be a major stumbling block to implementing DFT. The product itself has not changed, only its documentation within the manufacturing company.

Flattening out the bill of material is a major change in thinking for the traditional manufacturer. The objective is clear: As a company transitions to a Demand Flow manufacturing environment, it should move aggressively to simplify the bill of material. However, the company must understand that the simplification process often encounters internal resistance.

Phantom Bills of Material

Some traditional manufacturers are very reluctant to implement Demand Flow manufacturing and flatten the bill of material. These manufacturers attempt to evolve into Demand Flow manufacturing simply by reducing lot sizes (for example, "MRP II, Lot Size of 1"). Typically, they attempt to go to a "blow-throughout, make-on, or phantom" bill of material. These phantom subassemblies still appear on the higher-level subassembly's bill of material. However, the scheduling system (MRP II) ignores them and does not try to open up work orders to issue material for the phantom assemblies. This is an unacceptable and dangerous approach, given that all of the complex problems associated with engineering changes, effective dates, and documentation systems related to a multilevel bill of material remain.

In addition, flow-line design, DFT operation-sheet management, cycle-time calculations, work balancing, and backflush control are unnecessarily complicated by these subassemblies, whether or not they are called phantom. This approach to flattening the bill of material should be avoided.

Because the simplified bill of material is a key in the transition from traditional to world-class Demand Flow Technology, it immediately affects several traditional departments and organizations. Failure to embrace this key element of the implementation may be an indication of a lack of support, understanding, and commitment from management.

Multipurpose Bill of Material

The bill of material in Demand Flow manufacturing serves two purposes: it identifies the parts required to build a product, and it relieves the inventory of the parts/materials consumed when the product is completed. This flow-manufacturing technique is referred to as "backflush inventory control." The bill of material identifies this inventory-control information as a backflush location. Inventory control systems directly utilize the DFT bill of material to relieve the on-hand quantity of inventory from RIP. Thus, it is mandatory to have a correct bill of material to ensure accurate on-hand inventories in RIP. Many companies practice backflush control for their relief of in-process inventories. The Achilles heel of this technique has always been the accuracy of the bill of material that identifies the component parts consumed.

This issue cannot be overstated, given the importance of maintaining an accurate bill of material. Traditionally, attempts to compensate for an incorrect bill of material include additional transactions, such as miscellaneous issues or unplanned issues from a storeroom. Demand Flow manufacturing works on the assumption that if the product was built, the parts on the bill of material were consumed. If the bill of material is incorrect, inventory is consumed incorrectly and the wrong information is fed to the procurement system for component replenishment. From the purchasing standpoint, the bill of material determines which parts to buy and when they must be delivered.

Managing Engineering Change in Demand Flow Manufacturing

Another major difference between the bills of material in Demand Flow manufacturing and traditional bills of material is that with Demand Flow manufacturing, the engineering-change system can utilize "pending change" implementation techniques. This enables the Demand Flow manufacturer to approve an engineering change prior to assigning an effective date or a corresponding revision level. The new materials can be ordered, but the revision level is not assigned until the change is ready to be implemented. Properly implementing an engineering change with a traditional multilevel subassembly bill of material can be a coordination nightmare. Effectively, each level of the multilevel bill of material must coordinate with each lower level. Part-number changes to a lower-level component can cascade up to higher levels. From a configuration-control standpoint, multiple changes to a multilevel product—attempted at different times with different effective dates—can border on the impossible.

Traditional manufacturers typically attempt to use product deviations to try to stem the cascading effects of these changes. Most products must improve and change rapidly, and the volume of engineering changes can remain high for some products. With the single-level bill of material, engineering changes become simple and effective.

Any change that affects the form, fit, or function of a product must have an accompanying engineering-change-order (ECO) approved by design engineering. Process changes (which do not in any way affect product form, fit, or function) may follow a more streamlined system. However, form, fit, and function should not change. Design-engineering review may be appropriate, but it should take place quickly. Process changes are implemented in a day or less.

Backflush Effectivity

When engineering indicates that a new part is introduced into a product via an ECO, the change is assigned an engineering-change number and approved through normal approval channels. Upon ECO approval, the new part that needs to be procured is listed in the system as a "pending change" with an estimated delivery date. The buyer then procures the new part that is called out on the engineering change. However, this pending change does not affect the revision level of the product, nor does the current bill-of-material information change. The engineering-change order is "pending" until the required material is received and the change is actually implemented. If a different engineering change is approved prior to receipt of the new part associated with the first change, and this second engineering-change part is immediately available, then the second engineering change is assigned the next revision level when it is implemented. The first engineering change is still considered pending until its part arrives. In Demand Flow manufacturing, when the manufacturer is ready to implement the change, the first unit going through the process with the change is tagged with the engineering-change number that authorized the change.

When this tagged unit reaches the point of backflush, only then does the higher-level revision get incorporated. The revision level is assigned, the bill of material is updated, the pending change is removed, and the material associated with the bill of material is correctly backflushed. This technique is known as "backflush effectivity." It requires the changed unit to be produced sequentially in a flow process—the preferred method in Demand Flow manufacturing. If another change is to be made, the same process is followed. Revision levels are assigned sequentially as the change is actually implemented.

This approach reduces the problems commonly associated with dated ECO changes attached to revision identifications that are "pending" in the schedulized system (MRP II). In Demand Flow manufacturing, the date assigned to the ECO is an estimated date, and the actual implementation date is the date the change was made on the production floor.

Changing Form, Fit, or Function

Demand Flow manufacturing's flat bill of material facilitates immediate and future engineering changes in a way that is easier, faster, and more accurate than that of the more complex multilevel bill of material. This is because the bill of material does not contain the traditional multilevel interrelationships and the corresponding domino effect of changes. If an engineering change affects the form, fit, or function of a product, the component and higher-level-product part numbers must be changed. If an engineering change does not affect form, fit, or function, that change is simply a revision change to the part number, and its higher-level product is not affected. Part number revisions never appear on a kanban. An ECO always documents revision changes, and on-hand or in-process material should never have a rework disposition.

Planning Bill of Material

If a product has several hundred different options, the possible number of unique products to document, plan, build, and cost could become unwieldy. In cases of products with multiple options, a planning bill of material identifies the common part for each product. Individual bills of material for the options are backflushed separately. The product could be configured at order entry to include the base product and the specific options for the sales order. The sales order, which is configured, takes on a unique number identity. The sales order (sales order number) is back-flushed at completion. This basic-product bill of material is a "pile of parts," with an independent bill of material for each option.

Plethora of Options

Consider the auto industry as an example. Each car comes with options for dozens of different colors, sound systems, interior fabrics, interior colors, wheels and tires, and so forth. A flat bill of material for each possible combination proves unwieldy and unnecessarily complicated. This approach can be dangerous and warrants caution. On the other hand, products with 20 or 25 possible options aren't sufficiently complex to force a planning bill of material. A planning bill of material is only required with a large number of options or combinations of options.

Independent Processes and Divisions

If a company produces a product in a flow process, with feeder lines attached and pulled into the final product, its product bill of material should be a pile of purchased parts. If a company has centralized product-design engineering, along with multiple independent divisions, this creates a multi-level bill of material. This is the case with divisions, which are managed independently and controlled by a central product-design group. The divisions have independent finance, planning, and management and, indeed, can make a profit on products sent to other divisions. In these cases, design engineering should structure a part number for the assembly produced in one division and sold to another internal plant (see Figure 8.5).

For example, consider an independent feeder division that manufactures printed circuit boards and sends them to two independent systems plants that are internal divisions of the corporation. With one central product design, a corporate two-level or even three-level bill of material results. To the printed-circuit-board division, the bill of material is a pile of components to make the assembly. To each of the system divisions, one of the purchased parts they procure is a printed-circuit-board assembly, which just happens to come from a supplier division. The two system divisions have a flat bill of material as well. To the corporation, a multilevel bill of material has been designed. This is a perfectly acceptable scenario.

Demand Flow Manufacturing Documentation

Documentation used by the Demand Flow manufacturer differs from the traditional documentation systems used today. As the bill of material is compressed, subassembly drawings are no longer required. Visual, graphic TQC operational sheets replace traditional assembly instructions loaded with text that most employees avoid reading. The design engineering team continues to control engineering changes that affect form, fit, or function of a product. An operational-sheet ID (OP ID) is related to the components on the bill of material. This enables engineering to list a "part number/where used" on TQC OP IDs. When the engineering change is approved, the process documents that need to be modified are already being identified.

The manufacturing engineering team controls process changes that do not affect form, fit, or function of a product. These changes tend to be frequent and quick to implement. TQC operational sheets tell the operator what to do and what to inspect or validate.

The total quality check (TQC) and employee-involvement processes identify causes for defects on current products. While design engineering must continue to develop new products, it also must place a renewed emphasis on existing products to correct design problems. While not all changes can be made immediately, design engineering must become an integral part of the continuous improvement process. The mentality of "throwing" products over the fence to manufacturing and proceeding with the next new product should be avoided.

Bringing Quality to the Process Level

The Demand Flow manufacturer documents the production method used in the process one time only, based on the highest desired rate and shortest cycle time possible. The calculated work content is an operation, and the documentation of the work is a DFT operational sheet. If an engineering change order is approved, an operational sheet may have to be changed—but only those operational sheets affected by the change order are revised. The Demand Flow process is designed and documented one time. However, changes to DFT operational sheets occur frequently, based on the engineering-change orders and the employee-improvement programs. Verification and TQC points must also be clearly colored on the operational sheet.

The Demand Flow manufacturer embeds quality into the process, rather than relying on external inspection. The TQC operational sheets guide the operator to these points. Employees may verify their own work, but TQC points validate work performed at a prior operation. If a TQC point is covered up and cannot be verified, this is identified as an acceptable-quality-level (AQL) product containing a design for defect that is assigned to design engineering.

TQC Operational Sheets

TQC operational sheets are used to communicate graphically to the production employees precisely what they are to do in terms of work content, verification, and total quality control (see Figures 8.5 and 8.6).

Each sheet is created directly from the TQC sequence of events. The sheets should have minimal text. Unit illustrations dominate the sheets, clearly showing the work to be done at that station, how it is to be done, and what is to be verified against work performed at the operation. Work content is coded yellow, work to verify is coded blue, and total-quality-control work is coded red.

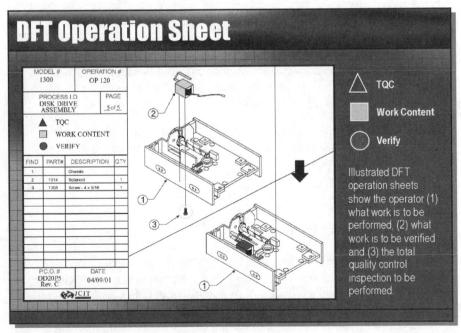

Figure 8.5 — DFT Operation Sheet

The find number, part number, description, and quantity of each part are listed, as is the process point, product number, and operation number. The sheets eliminate guesswork and exclude red lines, alterations, and modifications. If changes are to be made, the sheets must be redone. The operator should not need to interpret a sheet full of pencil scratches wondering which change was the latest and who authorized it.

Demand Flow Manufacturing Systems Integration

Formal computer tools are available to the world-class manufacturer to create an effective documentation network. The computer-aided-design (CAD) system can interface with the pile-of-parts bill-of-material and operational-sheet IDs to download component drawings and specifications, as well as quality specifications. This enables the manufacturing and quality engineer to use the design-specification documentation as an aid to creating bills of material and TQC operational sheets quickly. However, the creation and maintenance of TQC operational sheets should not be attempted in CAD.

Far more cost-effective PC-based graphic applications are readily available. PC-based LAN networks can also be effective. This network should interface with the bill of material to find and update the appropriate operational sheet when an engineering change is generated. The network links graphic software to hardcopy output devices with the objective of making an operational sheet change and getting it to the production floor in 30 minutes or less.

Another possible method is to interface to the manufacturing-system mainframe. PC networking technology can be incorporated quickly, without large investments in hardware or software.

Only One Bill of Material

There is only one bill of material in a world-class manufacturing environment—not a separate design bill of material and manufacturing bill-of-material. The difficulty posed by the multiple bills of material is that it is impossible to maintain consistent information. While the pile-of-parts information and specifications are clearly under design-engineering control, other elements associated with the bill of material, such as backflush and TQC operational-sheet information, are maintained by manufacturing. Ideally, everyone shares one company-wide bill of material that is 100 percent accurate and always consistent.

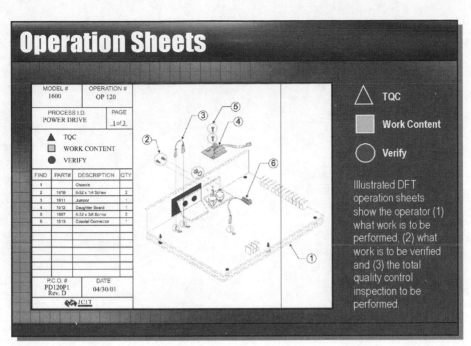

Figure 8.6 —- Operation Sheets

Chapter 9

DFT Procurement Forecasting and Quality Source Management

Unfortunately, some manufacturing companies begin their Demand Flow Technology (DFT) implementation program with their suppliers and procurement process. These companies mistakenly believe that DFT creates an inventory-reduction program that forces suppliers, rather than the consuming manufacturer, to carry inventory. The program sounds attractive on the surface: Drive down inventory investment via more frequent deliveries, reduce costs via negotiated agreements based on higher volume, and receive-higher quality material based on tightening specifications and eliminating inspection. This is a highly attractive, alluring prospect. It is also a very dangerous trap.

Suppliers Are Not Warehouses

Demand Flow manufacturing should never start with suppliers, and effective procurement programs should never drive suppliers beyond the capabilities of the consuming manufacturer's own internal process. It is of little benefit to insist on daily deliveries from suppliers while the manufacturing process continues to schedule, pick kits, and build multi-level subassemblies. Transportation costs will rise as deliveries increase.

Without a freight program geared toward reducing total transportation costs, overall product costs will also increase. Reducing lot sizes and increasing deliveries can increase storeroom transactions and costs. Eliminating incoming and/or in-process inspectors without first ensuring a quality part and process at the supplier results in higher quantities of defective purchased materials in the product-assembly process. As a result, adverse scrap and rework costs will go up.

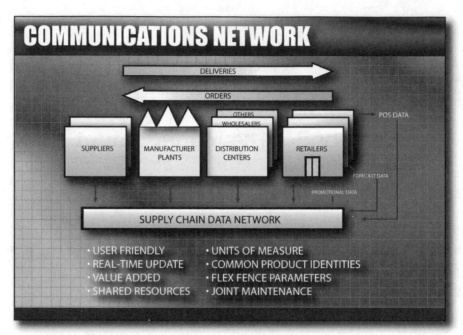

Figure 9.1 — Communications Network

Closer Relationship

Demand Flow manufacturing procurement programs can be effective only as part of an integrated company-wide implementation that views material suppliers as an extension of the Demand Flow production process.

The basic premise of the buyer/supplier relationship must shift from an antagonistic, price-based engagement to a cooperative relationship based on mutual benefit.

Personnel in the procurement department find their roles changing as well. Buyers spend far less time order launching and expediting and more time on flexibility management, contract negotiation, and the packaging and transportation aspects of the procurement process.

Suppliers Need Education

Just as the internal DFT implementation begins with defining the TQC sequence of events, defining operation-cycle times and points of kanban, and documenting the process, so does the process of educating suppliers. Suppliers must gain a basic understanding of flow strategies and technology. Suppliers must also understand that, although the company is not operating under Demand Flow manufacturing today, it is moving swiftly toward becoming a Demand Flow manufacturer.

One of the most misunderstood and misused aspects of DFT is supplier management. Many suppliers are extremely cautious about flow (JIT) manufacturing because they have the erroneous impression that flow manufacturing requires suppliers to store buyers' inventory. Suppliers see that they must make frequent, sometimes daily, deliveries, and produce 100 percent defect-free materials while reducing costs.

Demand Flow manufacturing extends the in-process quality and material-pull techniques back to the source for higher-quality and lower-cost materials. The supplier becomes an extension of the manufacturing process in a relationship that tends to be more mutually advantageous. Many of the guidelines for the relationship between the Demand Flow manufacturer and supplier are predicated on common goals, mutual trust, and mutual advantage.

Use Supplier Expertise

Demand Flow companies should make every effort to design parts that are easily produced. They should select standardized components, making sure to eliminate variances between supplier specifications and manufacturer requirements. Design engineering should try to use existing preferred components wherever possible.

Suppliers should participate in part selection and design. In many respects, no one knows better how to select components and meet specifications—or is better qualified for recommending modifications—than the supplier. Knowledgeable design engineers and procurement buyers understand that suppliers are often the best source of expertise in component selection. Excellent partnerships with key suppliers benefit both partners and often reduce sourcing costs. The world-class manufacturer seeks a consistent approach to standardizing and meeting specifications for the material used in all phases of the design cycle and the procurement process.

Supplier Cooperation

Supplier relationships in Demand Flow manufacturing don't involve pressuring the supplier or trying to increase the supplier's costs. The objective is not to force the supplier to build a plant near the manufacturer's operation or have the supplier store the manufacturer's inventory. The new relationship with the supplier starts very slowly. The Demand Flow manufacturer must first address its own internal process and achieve greater than 90 percent internal linearity before pulling suppliers directly into the process.

One best practice is to hold a "supplier day" at the Demand Flow manufacturer's plant. The manufacturer can brief suppliers on Demand Flow processes and strategies. The companies can exchange data, implementation plans, and schedules. The manufacturer can also emphasize that supplier conversion to Demand Flow manufacturing is preferred but not mandatory. Many Demand Flow manufacturers have gained excellent support from their suppliers, and the suppliers have improved their business performance and profitability via a mutual strategy of pursuing DFT implementation in their facilities.

Promote Understanding

It is important that suppliers understand the Demand Flow manufacturing philosophy and associated technology used in a manufacturer's plant and processes. This assists the supplier in understanding the manufacturer's internal needs, designs, and specifications.

The manufacturer must explain its strategies to the entire supplier base, focusing initially on the first wave of parts that are involved. Suppliers should understand that reducing the total number of suppliers means greater volume for the remaining suppliers. The explanation of flow strategies to the supplier includes:

- Involvement in the design process
- Contract/blanket purchase orders and implementation of quantity releases
- Flexibility requirements and management
- Packaging strategies
- Transportation strategies
- Single-sourcing goals
- Quality-at-the-source techniques
- Demand Flow manufacturing forecasting contracts
- Use of advanced communications
- General strategies

Instead of issuing traditional single purchase orders, the manufacturer and suppliers negotiate a DFT contract and requirement releases made against the contract. The supplier and manufacturer use computer-driven electronic data interchanges to enable simple releases, with less paper-work, to support more frequent deliveries. The use of fax communications to pull mutually agreed-on kanban quantities into the manufacturer's process is another popular success. While the frequency of deliveries will increase, a corresponding increase in product costs is not acceptable. The manufacturer focuses on reducing the quantity of parts in the bill of material and reducing the quantity of part numbers. If the Demand Flow manufacturer started with 5,000 suppliers, it should concentrate on reducing that number to 2,000, then to 500. Packaging is reviewed and possibly redesigned with an eye toward sizing for RIP quantities and reducing handling damage and costs. Quality parts, onetime delivery, and lower costs are the primary objectives.

Advanced communication tools help lower costs without increasing paperwork. Transportation networks also receive special attention. The ultimate goal is to trim the number of suppliers of the same part and commodity down to one preferred supplier.

Why Single-Source Suppliers?

Demand Flow manufacturing's emphasis on single-source suppliers should not be confused with the sometimes scary practice (often employed by regulatory agencies) of using sole-source suppliers. Sole-source suppliers are the only ones in the world that make or supply a particular part. If a company wants or needs that part, it must go to those suppliers and meet their terms. In contrast, single-source suppliers are those selected by the manufacturer or customer to be the exclusive supplier for a particular part. This exclusive selection is the decision of the selection team and buyer, and it offers several advantages. The choice of doing business exclusively with one supplier of a particular part differs from a sole-source approach—because there are other suppliers willing and able to provide the particular part.

Highly Advantageous

There are several advantages to converting to single-source suppliers. In single sourcing, suppliers have a negotiated contract. The challenges of qualifying a second source and the potential for higher costs of ensuring quality from the backup supplier offset any disadvantages. It often proves worthwhile to pay tooling costs for a second supplier, even if the need for the supplier is unlikely. Whether or not to qualify a second supplier is a business decision in which added costs must be balanced against possible long lead times and quality problems should issues develop with the single-source supplier. With single-source suppliers, purchased volumes should not amount to more than 30 percent of the supplier's business. An amount greater than 30 percent puts the supplier on shaky ground if the manufacturer's business drops. If such a supplier goes out of business, supply problems for the Demand Flow manufacturer quickly develop. Manufacturers that account for more than 30 percent of a company or division of a major corporation are the exception to this rule.

Better Control, Bigger Slice

The advantages of using a single-source supplier include:

- Being able to focus on that one source
- Consolidating volume
- Consolidating by commodity
- Maintaining better control
- Monitoring quality and supplier performance

Also, a supplier that gets the whole loaf instead of a slice is much more inclined to keep the buyer happy. The buyer has more time to focus on a smaller number of suppliers and can better monitor performance and contract flexibility.

Focus on Total Cost

In Demand Flow manufacturing, the emphasis shifts from the traditional, standard material cost of a purchased item toward the total cost of the part. Other costs in addition to the standard material cost must be considered. The costs of quality, transportation, packaging, setup, and carrying inventory are elements of total cost. The supplier with the lowest standard cost may not have the lowest total cost due to high rework, scrap, or transportation costs.

Initial Focus

Converting suppliers to Demand Flow Technology is a gradual and deliberate process. During the inspection of their premises, manufacturers should inquire into union affiliations, customer turnover rates, the number of new production employees, and any manufacturing process that has had multiple managers in the past six months.

The buyer inspects things that might interrupt the flow of quality goods to his plant. In qualifying and certifying the supplier and its processes, the buyer investigates the financial stability of the supplier, neatness and orderliness of the plant, and plant safety. Questions to ask include:

*   Is there a logical physical flow to their product?
*   How does the supplier measure and control the production flow?
*   How aware are the employees of current production techniques?
*   How do the supplier's financial ratios compare to industry norms?
*   How stable is the management team?
*   Does the quality emphasis occur at the process source rathe than final product inspection?
*   Does the supplier understand process-capability techniques?
*   Can the supplier prove its process capability statistically, and does it understand the variables related to its own process?

As a part of the certification process, Pareto analysis and control charts are set up to monitor parts at the supplier location; data is sent to the buyer's quality team with a closed-loop corrective-action program; and arrangements are made for periodic audits of the supplier's facility, processes, and materials. While the supplier becomes an extension of the buyer's production process, the buyer's quality development team becomes an extension of the supplier's process.

Design Process Modifications

Product design is crucial in achieving world-class manufacturing excellence. However, designers tend to proliferate part numbers. If several existing parts could be used for a new product, designers are likely to create a new one. Often, design engineering does not access the preferred part and supplier databases that can point them to preferred part numbers and preferred single-source suppliers. Computer-aided-design (CAD) tools and the network links to the resident manufacturing system must be readily available to aid the designer in part standardization and selection.

Supplier Involvement

The supplier should participate in the design process as early as possible. The Demand Flow manufacturer gives the supplier a preliminary set of purchase specifications that include required and optional dimensions, tolerances, and performance attributes. The required specifications, attributes, tolerances, materials, processes, and costs must be met. Other attributes of a component can have optional parameters in which the supplier can offer trade-offs. For example, the supplier can produce a two-percent tolerance component using alternate materials for a $10,000 tooling cost, or $0.10 per unit. Or, they can produce a one-percent tolerance device using an alternate technology for $0.15 per unit and $100,000 tooling costs. The buyer/engineer can select the appropriate technology depending on the required device specification.

The Demand Flow manufacturer must be careful not to "over specify" material and obtain a component that is far better but more costly than the specifications require.

Engineering should be encouraged to design existing materials/parts into new products wherever possible. At the conclusion of the discussions among design, purchasing, manufacturing engineering, and the supplier, the agreement should cover each part specification. At this point, there should be no confusion over the definition of the quality criteria for the part and no "specification disconnect" between buyer expectations and supplier capabilities.

Developing a Contract

Ultimately, a DFT contract will cover as many different part numbers as possible. Initially, the manufacturer might target the highest dollar volume part (calculated by multiplying annual usage by unit cost). Drafting the contract follows supplier selection. The qualifying, selecting, and contracting process continues with suppliers, part by part.

Other priorities for contract selection include:

- Space-intensive parts
- Material with tight quality specifications
- Parts critical to the product
- Parts involving high transportation costs
- The impact of inventory carrying costs
- Contract elements

Several elements are essential to a successful DFT contract.
Under the quality provision, the suppliers agree to supply material
per specifications, to monitor and control their process as required to
ensure compliance to parameters and acceptance criteria, and to share
information regarding the control and monitoring of their process.
Changes in the supplier's process are agreed on in writing if they affect
form, fit, or function of the item produced. The contract includes:

- Stipulation that all amendments must be in writing
- Description and mutually agreed-on specifications with reference
  to critical parameters and acceptance criteria
- Total quantity over the contract life
- Beginning and ending dates of the contract
- The shared forecast and flexible lead time

Negotiated Flexibility and Forecast

A typical DFT contract clause might include provisions such as:
"Buyer will provide a rolling forecast equal in length to the buyer's
planning horizon, updated each week with windows for adjusting
positive and negative variances to forecasted quantities. Seller's
lead time for shipment shall be one week after receipt of the release."
This lead time is a consideration only at the outset of the contract.
One of the reasons suppliers' lead times are so long in traditional
manufacturing is that suppliers have no real visibility beyond their
lead time, because they have no idea what the next purchase order
will bring. Demand Flow manufacturing provides suppliers with the
certainty of additional business and the flexibility needed to fulfill it.

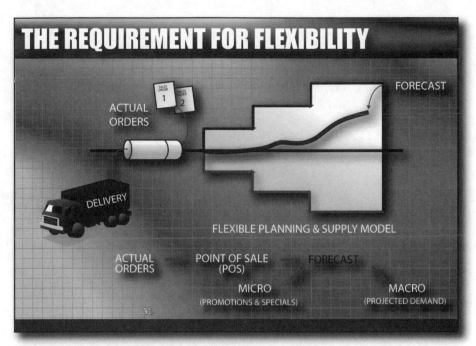

THE REQUIREMENT FOR FLEXIBILITY

FORECAST

ACTUAL ORDERS

SALES ORDER 1
SALES ORDER 2

DELIVERY

FLEXIBLE PLANNING & SUPPLY MODEL

ACTUAL ORDERS

POINT OF SALE (POS)

FORECAST

MICRO
(PROMOTIONS & SPECIALS)

MACRO
(PROJECTED DEMAND)

Figure 9.2 — The Requirement for Flexibility

Flexibility Grows

Consider a 24-month contract for one million parts. The contract would probably call for the release of 10,000 parts per week for the first four weeks before falling into the following flexibility-time fence: releases of between 9,000 and 11,000 each week for weeks four to nine; releases of between 8,000 and 12,000 each week for weeks nine to thirteen; and releases of between 7,000 and 13,000 thereafter for the duration of the contract. If the Demand Flow manufacturers do not use the flexibility in the contract, they potentially face excess inventory or a major shortage. Depending on actual demand, the Demand Flow manufacturer's needs could fall into the low end of the contract's flexibility (around 7,000), or increase to 13,000. The actual release quantities may be kept stable at 10,000 each and every week or respond to a demand-pull that reflects the latest volume required each week. The contract should include provisions for excellent communications if the pull quantity exceeds the upper flex limit negotiated.

Maintain Flexibility

Sales must tightly synchronize with suppliers. In signing the contract, the flow manufacturer commits to buying a percentage of the one million parts over the life of the contract. The commitment is part of the contract.

However, two stipulations can ease the situation if business conditions change:

1. Provide that the Demand Flow manufacturer can extend the contract for a reasonable time to consume the contracted quantity.
2. Negotiate a conciliation charge if the quantity commitment is not fulfilled.

Release provisions of the contract should stipulate that multiple releases will take place with the following variables to be determined: quantities, shipping dates, shipping method, carrier, destination, release mode, allowable variances from forecast and the quality requirement windows, and frequency of shipments. DFT contracts should avoid provisions for material with only short-term use or material that is likely to become obsolete. A careful check with design engineering should be made before contract negotiations begin.

DFT Supplier Communications

Stability in forecasts and the negotiated levels of flexibility are critical for success. It is impractical to request that the supplier react to changing contract release quantities or to allow frequent changes to the flexible limits. The successful contract provides the supplier the visibility for planning the capacity to meet the reasonable expectation of the forecast. Both supplier and manufacturer understand that the parameters of flexibility as well as the volumes potentially change over the life of the contract as business dictates. The contract should include agreed-on rules to accommodate these fluctuations. A typical example of a communication of demand forecasts for a purchased component is shown in Figure 9.3.

Notice that the purchase lead time is four weeks and that communication occurs weekly. The forecast is displayed from the last communication (one week ago) and the current week. Open releases are displayed to ensure that both parties are aware of the status of incoming quantities from the manufacturer's viewpoint. This allows quick resolution of any misunderstandings.

A tolerance range for quantity flexibility (based on flex percentages that increase as visibility extends) surrounds the first few weeks of the horizon, beginning at the end of the supplier lead time. These percentages are applied against a projected period-based volume. This period is equal to the frequency of the communication update (in this example, one week). The flexible percentages and the length of each flex fence are based on negotiations between the manufacturer and the supplier. The values are suited to the projected fluctuation of the components usage and the capacity capability of the supplier.

Figure 9.3 — Typical example of a communication of demand forecasts

Part Number 921317-01

Lead Time .... 20 Days

Period-Base Volume 200

Flex Percentage .... 10% | 20% | 30%

Flex Range (Weeks) .... 2 | 4 | 8

	wk 2	wk 4	wk 6	wk 8	wk 10	wk 12	wk 14
Open Purchase Orders	180	200					
Old Total Demand			210	200	245	230	200
New Forecast			210	235	210	230	200
High Flex Quantities			210	220	220	230	230
Low Flex Quantities			190	180	180	170	170
Violation?				YES			

As this example readily shows, released orders are expected for delivery in the next four weeks. The forecast is allowed to fluctuate, but when such fluctuations violate the flex tolerance, a prompt will initiate communication between the manufacturer and the supplier. This communication is critical and allows a negotiated settlement of the situation. The supplier may be capable of supplying the forecasted quantity at that time. This is the supplier's choice, but if it is not possible, then arrangements are made to resolve the problem. The key here is that arrangements are made in advance and allow a resolution without expediting shipments at the last minute.

Continual communication is vital to the success of the process, both to resolve issues as discussed above and to negotiate any required changes to the flex percentages, lead time, period-based volume, and other criteria for the planned supply to meet the manufacturer's needs. In DFT customer/supplier relationships, the supplier is still dedicated to serving the customer.

Contract Freight

It is essential for the Demand Flow manufacturer to maintain control of shipping. The DFT contract should state that shipping is FOB point of origin and that the buyer specifies shipping method and carrier.
The supplier is liable for incremental costs incurred when using an unspecified method or carrier. The buyer should have the right to accept or reject late and early shipments, as well as shipments in excess of the release or pull quantity. The contract should require the supplier to immediately notify the manufacturer of delays, and then complete the shipment via the fastest means—with increased cost paid by the supplier. In a flow system, early shipments can be just as bad as late shipments, due to the lack of a storeroom and the risks posed to the RIP-inventory system.

Inspection

Initially, the Demand Flow manufacturer inspects the supplier's material when it is received, even with a DFT contract in place. Depending on the supplier's quality history, the manufacturer may need time to become comfortable with the supplier's quality conformance before eliminating inspections. The number of inspections gradually decreases, and lots are skipped as the effort moves toward completely eliminating incoming inspections. If the manufacturer eliminates inspections too quickly, bad parts go undetected, exposing the Demand Flow manufacturer to increased rework and scrap costs. The DFT contract should stipulate that the Demand Flow manufacturer has the right to inspect the supplier's plant and production process. The contract should also specify the flow manufacturer's right to randomly audit the supplier's process. The manufacturer should assist the supplier in establishing comprehensive in-process quality techniques, in favor of parts inspection and sorting. Inspections of the supplier's process should initially occur once every few days and gradually decrease to once a month as the supplier's in-process quality improves. It may take a year or more to bring quality to the desired level, but such inspections minimize quality problems at the outset and bring the supplier to the quality level required as swiftly as possible.

Recourse for Noncompliance

A point system should be established to provide recourse in the event that the supplier fails to meet contract requirements to an appreciable degree. For example, early or late shipments might be one point. Early or late shipments that disrupt process might be ten points. By assigning points to each failure to perform on these and other contract provisions, the contract gives the manufacturer the right to cancel the agreement if the points reach a specified threshold in three months.

Invoicing/Price

Important guidelines in creating a DFT contract include the following:

- The DFT contract should specify invoice mode (such as written or electronic), plus applicable discounts and best time frame.

- A specific price should be quoted for the life of the contract, though the price may be allowed to vary depending on the term. It may be unreasonable to assume that a price can be fixed for a five-year contract, for example.

- Develop separate pricing for operation or process that may be deleted within the term of the contract to avoid repricing upon deletion.

- Allow for reductions in the price during the term of the agreement if the seller can realistically be expected to realize economic improvements through continual total quality control of the process.

- Define formulas for indexing against a particular commodity on unique raw-material prices.

The approach allowing price fluctuation further illustrates that in Demand Flow manufacturing, the supplier is an extension of the flow-manufacturing plant. If the supplier saves $1 per unit through a joint improvement in the process, the Demand Flow manufacturer should expect to receive a percentage of savings in reduced part costs. If, due to reasons beyond the supplier's control, the cost of the unit goes up, the Demand Flow manufacturer expects to share in that increase.

Manufacturers don't benefit from beating up the supplier or seeking an unrealistic and arbitrary price reduction for a specified period without assisting in improving the production processes.

Advantages to Supplier

There are many advantages to being a flow supplier to a Demand Flow manufacturer:

- The suppliers receive a longer forecast or horizon of visibility, which facilitates better planning and reduces lead-time requirements.

- The forecast has a flexibility component that accommodates fluctuations in business.

- The suppliers receive frequent releases, and consistent shipments mean consistent cash flow.

- The Demand Flow manufacturer will consolidate invoices to some extent but will still pay its flow-manufacturing suppliers quickly. In fact, the manufacturer should promise suppliers that the manufacturer will not age supplier invoices.

- Setups occur frequently and become a consistent part of each supplier's process of improving quality and production.

- Frequent shipments allow the suppliers to carry smaller inventories.

- Suppliers receive immediate feedback on quality problems and the required process improvements.

- Single-source suppliers receive 100 percent of the manufacturer's business for the particular part or parts involved and the security of a long-term strategic relationship.

- Suppliers can better plan materials and plant capacity.

Better Cash Flow

The supplier's advantages are particularly clear when applied to different time frames. For example, consider 1,000 units priced at $100,000 required during a four-week period. If the units shipped every four weeks, cash flows every four weeks, and the quality exposure is on 1,000 units. The supplier and the manufacturer are both required to carry inventory. There may be only one setup every four weeks, a time frame that will not assure process familiarity, setup-reduction efficiency, or reduced-quality exposure. In contrast, consider the same usage on a weekly basis. The cash flow is weekly. Maximum quality exposure is 250 units. Maximum inventory carried by either company is 250 units. Efficient setups can be weekly instead of once every four weeks.

Transportation Strategy

Apart from the transportation provisions of the contract, cost-effective transportation is a goal of the Demand Flow manufacturer. In the Demand Flow model, transportation demands may increase, as well as the demand for better service and higher dependability, but no additional money should be spent on those objectives. There is no advantage to spending on transportation. Companies implementing DFT should consider the following transportation guidelines:

- Consolidate shipments geographically to qualify for the lowest rates.

- Replace ordering costs with transportation costs in the part period balancing formula to optimize costs when volume is not large enough for price breaks.

- Take advantage of the deregulation of the freight industry and negotiate better prices.

- Consolidate shipments with other companies and, where possible, coordinate trucking back hauls.

- Investigate rail and air transportation. No longer are the trucking companies always the least expensive mode of transportation.

- Consolidate freight volume to minimize costs. If the FOB point is factory, the contract may specify the use of the same carrier for inbound as well as outbound shipments.

- Explore all options and monitor industry regulation changes. Do not award business on price alone.

- Award long-term contracts. Negotiate a DFT contract with the transportation company similar to the contract with suppliers. Make sure quality and responsibility for parts damaged in transit are a part of the contract.

Packaging Strategies

Henry Ford is reputed to be the first person to recycle packaging in the United States. The story goes that he bought the engines for his Model T from the Dodge Brothers, and he asked them to drill three holes in specific places on the side of the wooden boxes used to ship the engines. They complied and later asked him why. The reason: Henry Ford used that panel of the box as the floorboard in the Model T, and the holes were for the pedals. Packaging does not always allow for that method of reuse, but it does reward the creative recycler.

For example, on a particular computer base, negotiations with the supplier required that the base ship with a specific pallet attached. When the base was received, the pallet was left attached. The manufacturer built the product on the base, and because the pallet wasn't removed, they traveled through the production process together. Upon completion, the shipping box was attached to the original base and pallet, and the completed unit was shipped in the same container that brought in the base component. That kind of utility eliminates both a trash problem and the cost of additional packaging.

Many Opportunities Presented

The DFT contract should contain the following packaging specifications:

- The supplier will be required to package material per specified standards with changes mutually agreed on and detailed in writing.

- Packaging should facilitate handling, counting, and storage.

- Packaging should afford maximum protection to ensure that the quality of the parts remains intact.

- Quantity per package and/or size of package should facilitate delivery to and use at the line in a kanban quantity.

- Trash problems or recycling should be considered in package design, as should electronic bar-coding techniques, which may streamline receiving.

Many DFT manufacturers have made arrangements with suppliers to introduce returnable containers for component parts. The container is designed to suit the component part, satisfy the quantity required, promote ease of handling, and maximize container use at the point of consumption on the flow line. The objective of packaging is to eliminate as many non-value-added steps as possible in unpacking, repackaging, counting, and processing trash. The design engineer, process engineer, and buyer should focus systematically on these topics and eliminate the added costs for these non-value-added steps.

Advanced Communication

The advanced communication systems available today can lower ordering costs. Consider the following example: forecasts and releases are trans-mitted via network; bar codes on incoming packaging facilitate receiving; invoices are also transmitted via network; payment occurs through an electronic-funds transfer. Speed is enhanced, with minimal paperwork.

Even with the elimination of traditional transactions, there is no better evidence than that provided by the Demand Flow manufacturing process—if you built the product, you must have had the parts. This topic is discussed further in Chapter 13.

Initial Purchasing Priorities

The Demand Flow production process is well underway before a contract receives any consideration. Before negotiating a DFT contract, the manufacturer's priorities include resolving major quality issues and assisting production in becoming linear, with the ultimate goal of delivering customer service at a 99 percent satisfaction level. Once the production process achieves linearity, the manufacturer can select primary suppliers to begin strategic relationships based on the following:

- Sole-source parts
- High-dollar parts
- Space-intensive parts
- Critical-specification parts
- Inventory carrying costs

Quality techniques and freight issues become top priorities. The manufacturer begins the process of defining optimal lot size, adjusting deliveries according to demand, and reducing safety stocks as quality and delivery problems are solved.

Second Phase

The second phase of contract negotiations may include:

- Additional packaging improvements (taking the parts directly to RIP)
- Electronic invoicing/order release
- Eliminating change orders and communicating releases against the contract.
- Biweekly deliveries
- Tightening allowable performance points
- Earlier supplier involvement in future product design changes

The corresponding supplier benefits include:

- An increase in volume due to the reduction of the supplier base
- Weekly forecasts rather than the estimates based on the previous month, quarter, or year
- Guaranteed quantities
- Improved quantity and delivery performance to a major customer

Example of Savings

One of the divisions of GTE that converted to Demand Flow Technology enjoyed improved delivery service on components while saving $800,000 per year on price and quality improvements. They found the source of defective units in a conflict in specifications between them and their supplier. They also found that bringing in components that were stored for several months in a standard cardboard box reduced the solder ability of the component's leads. Under the DFT contract, the supplier packaged components in electrostatic-discharge-sensitive packaging, and GTE drastically reduced storage time. The component defect rate went from three percent to 150 parts per million. Price per unit went from $.79 per unit in 1986 to $.65 per unit due to a doubling of volume and the elimination of secondary suppliers. Supplier-quoted lead time went from 20 weeks to 4 weeks.

One supplier received orders for 2.8 million units, rather than three suppliers providing 933,000 units each. GTE and its suppliers also made additional packaging, inspection, and other related process improvements. The bottom line was that GTE realized savings of $400,000 a year in price, $200,000 a year in rework costs, and $200,000 a year in incoming inspection costs.

These kinds of savings are not unusual in a flow manufacturer/supplier relationship. Other DFT manufacturers have achieved results that dwarf those of GTE.

Changing Role of the
Demand Flow Manufacturing Buyer

Today's traditional buyers spend most of their time on order launching and expediting. This includes reviewing the requirements for purchased material, determining which of several qualified suppliers will receive the order, phoning the supplier for confirmation, mailing a purchase-order document, and following up to ensure on-time delivery (or conducting a postmortem to find out why it was late). Typically, the original due date of the purchase order with its long lead time will differ significantly from the actual "need date" of the purchase order. After navigating multiple planning cycles, which can impact the due date of the purchase order, the buyer struggles to keep up with the "pull ins" and often has no time for the "push outs." Materials-planning reports and traditional management proliferate. Each time a planning cycle occurs, the traditional MRP II computer system recalculates all existing required dates and distributes messages on how to reschedule existing purchase orders.

Process Gyrations

Because the traditional planning cycle can cause wild variation in requirements, purchase-order due dates "bounce" from one date to another. This creates havoc for the buyer attempting to manage this process and for the supplier attempting to keep up with this variation. Often, buyers have so many messages due to the planning-system sensitivity that they cannot keep up.

Also, the traditional relationship between the production floor's requirements and the MRP II reports is somewhat tenuous. For example, six work orders or kits of parts in staging, requiring a quantity of six of the same part numbers, can't be released with just one received part. All six are needed. All material, once allocated to the six work orders, is frozen, and that material cannot be used on other products. Determining the requirements to "keep production going" can be very confusing for the buyer. Buyers often revert to manual "hot list" systems generated by manufacturing.

Purchase Orders Costly

The material that the traditional buyer procures has specific purchase-order line-item identity, quantity, and delivery-date definition. The average cost of creating a purchase order ranges from $50 to $200, depending on the company. This includes paperwork, postage and handling, and clerical and buyer time. It does not include expediting, quality, and shortage costs. Since the cost of the purchase order is so high, traditional buyers tend to group, or "lot size," material requirements and buy bigger batches to minimize ordering costs. The formula by which traditional manufacturers tend to balance inventory carrying costs and purchase-order-placement cost is known as "part period balancing" (PPB).

For example, if part number 17203 had a purchase-order cost of $100 and an average inventory of 1,000 units, a company would incur a standard cost of $1 and 10 percent inventory carrying cost per year (See Figure 9.4).

Figure 9.4 — PPB formula example

Procurement Cycle = $\dfrac{10\% \times 1{,}000 \times \$1}{\$100}$ = 1
     (PPB)

Order/Purchase Order Cost = $100

Inventory Carrying Cost per Year = 10%

1,000 (average inventory) x $1 (unit cost) = $1,000

The calculation in Figure 9.4 indicates that the part should be purchased once a year. The most the part could be turned is once a year, because it is purchased in 12-month quantities. This scenario demonstrates the problem with the traditional PPB formula. The order cost per purchase order is used to calculate how often to buy the part. If the cost of ordering were incurred only once a year via contract, and deliveries occurred more frequently, the inventory could also turn more frequently, without adding additional overhead. This is the objective of the DFT buyer.

It is highly recommended that DFT buyers plan as well as buy their own material. Material planning, through the single-level, bill-of-material explosion, will inform the buyer of the need for flexibility in the negotiated plan. If sales has increased the forecast, purchasing must advise the supplier of the increased requirements (assuming the requirements fall within the flexibility parameters). Figure 9.5 provides an example.

Figure 9.5 — Flexibility in the negotiated plan

Product: Model 797X

Negotiated Level: 100/week

Purchased Part: 92131701

Quantity Per on Bill of Material: 1

Delivery Cycle: Weekly (currently 100/week)

Sales has requested an increase for Model 797X, totaling 120/week beginning in week six, which is within the negotiated flex fences. Material planning would indicate the following:

	wk 1	wk 2	wk 3	wk 4	wk 5	wk 6	wk 7	wk 8
Part Number: **Model 797X**								
Gross Requirement	100	100	100	100	100	120	120	120
Scheduled Receipts								
On Hand	0	0	0	0	0	0	0	0
Net Requirements	100	100	100	100	100	120	120	120
Part Number: **921317-01**								
Gross Requirement	100	100	100	100	100	120	120	120
Scheduled Receipts	100	100	100	100	100	100	100	100
On Hand	0	0	0	0	0	0	0	0
Net Requirements	0	0	0	0	0	20	20	20

The supplier receives notification, preferably electronically, that beginning in week six, 20 additional units, totaling 120 units, are needed. No offset of the supplier requirements is necessary if the requested flexibility is within the negotiated contract parameters. The buyer may decide to carry some additional inventory temporarily to cover quality, delivery, or other problems. This inventory may also cover sales flexibility requirements. For example, assume the buyer is required to keep an additional 10 percent inventory for a critical part that is difficult to forecast. Using the previous example, the material plan would indicate the following:

	wk 1	wk 2	wk 3	wk 4	wk 5	wk 6	wk 7	wk 8
Part Number: Model 797X								
Gross Requirement	100	100	100	100	100	120	120	120
Scheduled Receipts								
On Hand	0	0	0	0	0	0	0	0
Net Requirements	100	100	100	100	100	120	120	120
Part Number: 921317-01								
Gross Requirement	100	100	100	100	100	120	120	120
Scheduled Receipts	110	110	110	110	110	110	110	110
On Hand	10	20	30	40	50	40	30	20
Net Requirements	0	0	0	0	0	0	0	0

Figure 9.6 — Keeping additional inventory for critical parts that are difficult to forecast testing the material flexibility of the contract. This may be an effective, although expensive, preferred strategy on low-dollar and low-volume parts. Buyers should avoid netting out traditional safety stock quantities in advance of the material planning. Safety stocks should remain visible even as they are eliminated.

Demand Flow Manufacturing
Initial Purchasing Goals

The buyer should focus on getting parts for DFT contracts and then manage flexibility. An initial goal of Demand Flow manufacturing purchasing should be to include 70 percent "Class A" parts on a DFT flexible contract. This would mean that a DFT flexible contract would initially cover only a small percentage of parts. However, over half of the annual investment in material would be positively affected. As the Demand Flow Technology implementation progresses, Class B parts can be scrutinized and evaluated for commodity contract consolidation.

If a company has 1,000 purchased active parts, the initial goal should be to have 35 Class A parts on contract by the end of the first phase. These parts should be linked to the Demand Flow manufacturing initiative. Daily deliveries can only begin when the company attains greater than 90 percent manufacturing linearity.

The DFT buyer should also negotiate contracts for transportation and begin to build freight networks. The buyer's goal will be to increase the number of deliveries without increasing transportation costs. Inexpensive Class C items can arrive in infrequent deliveries. Here, the PPB formula can determine delivery frequency. As transportation costs go down, delivery frequency can increase.

For Example:

Day 1

Transportation Cost/Delivery: $50

Inventory Carrying Cost: $1 (standard. cost) x 10% (ICC)

Usage = 100/week or $10

A weekly delivery would cost more per week ($50) than the cost of holding the inventory ($10). Bimonthly deliveries will become appropriate as networking and single source transportation drive down transportation costs. Eventually, weekly deliveries may become appropriate. Buyers, with the aid of the computer, need to manage the delivery cycle with a focus on total cost.

Demand Flow Manufacturing Buyer Evolution

Traditional manufacturing buyers spend about 50 percent of their time in order launching and expediting. The other 50 percent goes to clerical functions, negotiating prices, keeping production going, managing by report, rescheduling deliveries in and out, and attending material review board (MRB) meetings. The buyer in Demand Flow manufacturing assumes different responsibilities: managing flexibility; kanban management and pre-shortage analysis; deep involvement in DFT flexible contracts; transportation, packaging, and supplier management; planning and buying material with total cost emphasis; and assisting in building a linear process. Buyers can have a positive impact on the total cost of a product. Their objectives should include a two to three percent material cost reduction per year through the use of flow techniques applied to their supplier base, and through the education and assistance of their new "team" members.

Chapter 10
Total Employee Involvement

Why become a world-class manufacturer? What is the purpose?
Aren't other manufacturing technologies and philosophies acceptable?
The global competitive reasons for converting to world-class Demand
Flow manufacturing technology are outlined in the beginning of this book.
The individual company reasons for implementing Demand Flow
manufacturing are simple:

- Greater customer responsiveness
- Quality improvement
- Overall cost reduction
- Survival

Actually, many companies are becoming proficient at reducing costs.
However, they achieve many of those cost reductions by meeting
an arbitrary percentage cut to maintain profitability rather than through
productivity, quality, and process improvements. Demand Flow
manufacturing cost reductions are true improvements. Expected
improvements in cost control can be substantial and dramatic.
An average company's Demand Flow manufacturing results can
be as impressive as the actual statistics in Figure 10.1 shown.

CATEGORY	Company A	Company B
WIP Dollars	*Down 50%*	*Down 65%*
Floor Space	*Down 25%*	*Down 40%*
Scrap/Rework Cost	*Down 44%*	*Down 35%*
Mfg. Output = People	*Up 16%*	*Up 20%*
Labor Efficiency	*Up 20%*	*Up 31%*
Inventory Turnover	*From 3 to 21*	*From 7 to 26*
Cycle Time	*From 8 weeks to 80 minutes*	*From 4 weeks to 40 minutes*

Figure 10.1 — Actual Demand Flow manufacturing results for two companies

These results and goals are realistic for the manufacturer who implements Demand Flow Technology in an organized, uncompromising manner. Demand Flow manufacturing should never be sold as a work-force reduction program. While it is true that the Demand Flow manufacturer requires a significantly smaller number of resources in some areas, other areas require more resources. Due to Demand Flow manufacturing improvements, the Demand Flow manufacturer produces higher-quality and lower-cost products to entice sales to sell more products that the same work force can support without an increase in personnel. Retirement and attrition take care of many work-force adjustments. Support of the DFT implementation program from all levels is mandatory for a successful implementation. A detailed analysis of how best to use a work force and free it from traditional functions is an inherent part of the DFT implementation process.

Economic Success

Many companies began converting to Demand Flow manufacturing during the mid '80s and '90s, with a substantial increase in the mid to late '90s. The reason was—and is—simply to maintain competitiveness and even surpass global competitors. In the United States, much has been made of the jobs created over the last 20 years. However, more than 80 percent of those jobs were in the service sector and are now at risk of moving offshore as well. In many countries, the number of manufacturing jobs continues to erode. This is simply and categorically unnecessary. Demand Flow manufacturing is the technology and foundation that can create and sustain a healthy climate for manufacturing jobs—both in the United States and abroad.

Most Important Asset

As manufacturing technology evolves from traditional, labor-tracking, scheduling, batch mentality to Demand Flow Technology, the role of people in the technology changes as well. People are the most important asset of any company. In a Demand Flow manufacturing environment, the responsibilities and work content of many employees change. As the roles change, the organizations to support the people tend to change. Many times in traditional manufacturing, people are told what to do and how to do it. With DFT, this mindset must change to promote a participatory approach to problem solving that involves a wider team concept within all companies.

Communication Failures

The traditional manufacturing company consists of many levels of management. Information, goals, expectations, and philosophies are reinterpreted as information moves down through the various levels. The phenomenon is rarely intentional, but each level brings a unique perspective to events and uniquely interprets information. By the time the information reaches the people and level for which it was originally intended, it may bear little resemblance to the initial message. Direct exposure of the lower tiers of the organization to the upper echelon of the company is infrequent and formal in nature.

Furthermore, information moving from the bottom up suffers the same fate. Middle managers typically do not want to bother upper management with details and information from the ranks. The information becomes increasingly summarized as it moves up. The information also suffers from a phenomenon known as "filtering," in which information detrimental to the middle levels is hidden or buried in statistical gobbledygook. This is human nature, but it makes it very difficult to get the proper information to the proper decision makers in a timely fashion.

Surpassing a Level of Competence

Another phenomenon typical in many manufacturing companies is the "Peter Principle," which states that a person will be promoted to his or her level of incompetence. Management positions are highly valued in our country for obvious economic and status reasons. An individual may be an excellent engineer, even though he or she has not received sufficient training or experience to become a manager of engineers. The individual desires the promotion and must break into management to progress and receive increased compensation. When the employee achieves his goal and the inevitable promotion comes, the net effect to the company is the loss of a good engineer and the gain of an inexperienced manager. In DFT, the reward systems for both direct and salaried employees allow for increased compensation based on results.

Employees are not the Problem

From the 1980s and into the 21st century, production employees in the United States often take the blame for the decline of manufacturing in this country due to the demand for high wages and other compensation, often through unions. At the same time, reports of waning productivity further contribute to the loss of jobs. Compounding this assault on production employees is the frequent cry that quality workmanship is severely diminishing. But an examination of the underlying causes reveals that the root of the problem is management practices, defective organizations, and outdated compensation strategies.

Throughout these developments a flood of books, articles, and consultants on organizational development, management styles, and theories have attempted to analyze these human-resource challenges. As companies attempt to compete and survive, the phenomena discussed below become clear.

Flexibility, Participation Come of Age

Flexibility is a valued skill—for individuals and for companies. The age of specialization is over. No longer can a work force focus on a narrow range of skills. Markets and companies are changing too quickly.

Participative management is a concept whose time has come. As organi-zations become leaner in an effort to remain competitive, they must use the resources available. They need to harness the creative problem-solving abilities of their employees, as well as their technical and administrative skills. Participative management is the ability of people affected by a decision to participate in the decision-making process.

Involvement, Compensation Change

Employee involvement must become ingrained in manufacturing culture and integral to a company's way of doing business. Management must make a leap of faith and assume that the person who best knows how to do and improve an operation is the person who is actually doing it. It is essential not only to seek input from the operator but to act on that input as well. Often, the launch of an employee-involvement program is followed by a flood of employee recommendations, many of which are not investigated. Employees become disillusioned and stop participating. In most cases, a nonexistent employee-involvement program is better than a nonsupported employee-involvement program.

Compensation programs must keep pace with the changing responsibilities and organizational shifts. Pay must correspond to flexibility, team and orga-nizational performance, employee involvement, participation, and resulting improvements. The seniority, pay-for-title, or position-based pay programs common in industry today don't fare well in a highly competitive environ-ment with rapidly changing technologies.

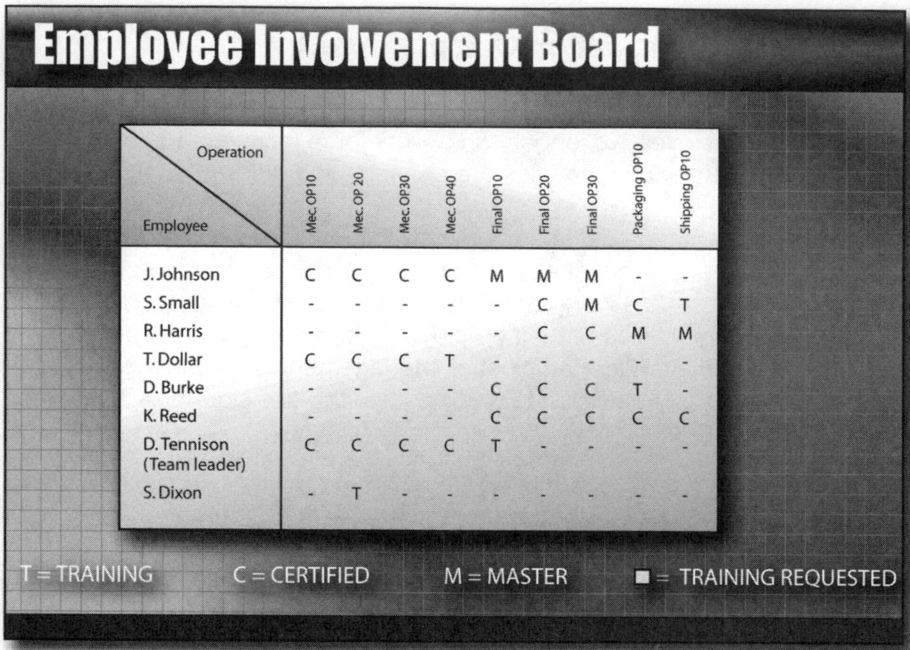

Employee Involvement Board

Operation / Employee	Mec. OP10	Mec. OP 20	Mec. OP30	Mec. OP40	Final OP10	Final OP20	Final OP30	Packaging OP10	Shipping OP10
J. Johnson	C	C	C	C	M	M	M	-	-
S. Small	-	-	-	-	-	C	M	C	T
R. Harris	-	-	-	-	-	C	C	M	M
T. Dollar	C	C	C	T	-	-	-	-	-
D. Burke	-	-	-	-	C	C	C	T	-
K. Reed	-	-	-	-	C	C	C	C	C
D. Tennison (Team leader)	C	C	C	C	T	-	-	-	-
S. Dixon	-	T	-	-	-	-	-	-	-

T = TRAINING   C = CERTIFIED   M = MASTER   ■ = TRAINING REQUESTED

Figure 10.2 — Employee Involvement Board

Make a Difference

In the 1980s, the Ford Motor Company completed a $6.3 billion turnaround. The company went from an annual loss of $3.1 billion at the beginning of the decade to a profit of $3.2 billion in 1987. When asked to what he attributed the dramatic turnaround, Ford's chairman, Donald Peterson, said, "A primary reason was people—participative management, and employee involvement." He defined employee involvement as "the soliciting of input of the employees on how to improve the process, and taking action on that input."

World-class manufacturing is accomplished through people—people play a more dramatic, extensive, and critical role than in traditional manufacturing methods. Employees receive more training and do more and different operations. They also have more and different responsibilities. That is why they are called flexible employees. They are paid for their flexibility, rather than their seniority.

Production employees are responsible for quality, and unlike in traditional scheduled manufacturing, the production employees can stop the line. Production employees can be trainees and trainers. They can move to leadership positions or to replenishing the kanban materials to aid material handling. In fact, they can fill any number of positions on a temporary basis to support the production environment.

Minimum One Up and One Down

Production employees in Demand Flow manufacturing must be able to work, at a minimum, "one up" and "one down." In other words, they must be able to do the operations on either side of them, which means they must be able to perform at least three different operations: their own, the one immediately before it in the process, and the one immediately after it in the process. An employee at the beginning of a process must learn his or her primary position, one position downstream and one position upstream.

An employee at the end of the process, in addition to learning his or her primary position and one position upstream, must learn the first operation in the next process or learn material handling.

If a production employee reaches for a unit to work on and there is no unit, that employee moves in the direction of the pull to work on a unit that supplies the empty station. The employees are not told to do this. It is an automatic response to the absence of units flowing to a station. Employees can help complete the units flowing to their station and then return to their station; or the next operator down the line moves down and takes a position in the then-vacant station. The process and movement of employees is that simple: an employee goes to pull a unit, no unit is there, and the employee flexes in the direction of pull.

Greater Production Flexibility

This flexibility is in marked contrast to traditional manufacturing, in which the employee remains at his idled position and perhaps reports the situation. With fewer people in the process, the observed operational cycle time increases, even though the work content does not change. Due to employee flexibility in the Demand Flow manufacturing system, a process can run with 50 percent of the employees absent. Although the observed total product cycle time increases and the volume of products decreases proportionally, the Demand Flow manufacturing process can run smoothly even if every other employee is absent. Conversely, as the need for products decreases, employees can be pulled and the process balances itself.

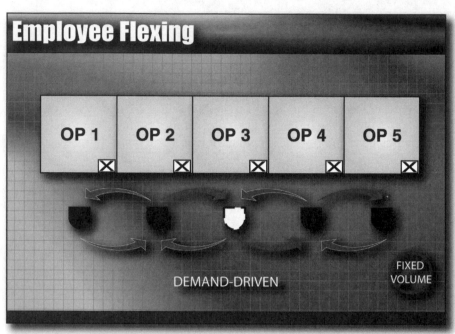

Figure 10.3 — Employee Flexing

In high-turnover situations, the tactic of "one up, one down" is even more beneficial. No matter how many holes in the process must be staffed by flexible employees, verification and total quality control continue, and work content does not change.

One up, one down is the minimum requirement to work in a Demand Flow manufacturing process. After attaining these skills, the employee may decide to pursue further flexibility standards, such as two up, two down, three up, three down, and so forth. Eventually, a few employees may attain certification in all process operations, while others choose to stay at the minimum level of one up, one down. Employees usually pursue flexibility horizontally and vertically (for example, doing several different assembly operations and doing assembly, testing, and machine trouble-shooting).

Pay for Skill

In DFT companies, employees are paid on the basis of individual flexibility, individual employee involvement, knowledge, skill, and teamwork. These, in turn, prompt major changes in the operation of the human-resources department. Significantly, an employee receives pay for knowing an operation, even when he or she may not perform it every day. Points are applied to different operations (positions). An employee can be training in a position, certified in a position to do the work without assistance or additional training, or a master in a position (one who has reached the highest level in that position, with a track record of quality work and the capability to train others). Although one up, one down represents the minimum requirement, many employees achieve more. It is not unusual for an employee to be a master in a few positions and certified in a few other positions. It is also not unusual for a heavily cross-trained and widely skilled employee to have higher pay than the team leader, with points varying per position, depending on the type of work content at the operation and as employees progress from trainee to certified to master status.

Programs

DFT programs have four elements of pay:

1. Base wage
2. Flexibility skill points
3. Team and organizational incentives
4. Compensation for employee-involvement program activities

The base wage of an employee is based on the existing market value for a given position in the plant's geographic region. This base wage is the average market value of any particular skill. Depending on the current employment rate in the area of the country at the time, companies have chosen to bring new employees into the Demand Flow manufacturing environment at 90 percent of market value for base pay. Yearly cost-of-living adjustments may apply to the base. This base pay may range anywhere from 50 to 100 percent of an employee's total compensation package. A new employee may therefore begin with less pay than a similar position in a traditional environment.

Team leaders face the challenge of providing the training necessary to new employees to quickly bring their knowledge and flexibility to the level required to gain a competitive salary increase. This process should require one to three months. If the new employee objects to this arrangement, he or she is probably not the type of employee suited to the Demand Flow manufacturing environment. The message is clear and fair: the company is looking for flexible employees who desire to work in a specific type of environment. The compensation potential with the combination of individual, team, and organizational incentives is always higher in a Demand Flow manufacturing environment. Labor costs per employee—the smallest element of total product cost—may increase somewhat, but they are offset by Demand Flow manufacturing's requirements for fewer employees. Decreases in material and other overhead costs (the predominant product-cost elements) overshadow the reduction in labor costs.

Flexibility

After an initial training period, all employees in a DFT environment are required to be certified in a minimum of three positions. Employees must know their primary positions, as well as one position up and one position down from their primary. This is required for several reasons. Because employees must verify the previous work content sent to them, they must be familiar with the work content of those positions. They must also be aware of the following position's work content, where that operator verifies their work. Also, management runs various flow lines at different rates. Employees are inserted in or pulled from a line based on the current rate of the lines. Each employee does the work and quality defined by the TQC operational sheets in that operation. As rates decrease, labor and machine requirements may decrease, but the designed work content and quality criteria at each operation do not change.

Employees Flex to the Need

Employee flexibility allows management to adjust the volume of products produced without changing the operational work, quality criteria, or process takt time (operational cycle time) of an operation. It also allows management to move employees to alternate operations without mass retraining efforts. The pull process requires flexible employees. Flexible employees are allowed to fill their in-process kanban (IPK) and complete the unit at their station or machine. At this point, their demand is satisfied and they must flex. Given these simple rules, only two situations can occur. In the first situation, a unit produced at the operation cannot enter the downstream IPK because a unit is already in the IPK—a situation downstream is inhibiting the flow of products to the customer. The flexible employee flexes downstream. They then assist at that position until a unit is completed at the downstream operation. In the second situation, a unit is built and placed in the downstream IPK. The employee then looks to pull the next unit from the upstream IPK and build a second unit to maintain the flow. In this situation, there is no second unit in the upstream IPK to pull into the operation— indicating an issue upstream that is inhibiting the flow of product through the line. The employee flexes upstream and assists at that position until the unit is completed at the upstream operation.

The flow line naturally rebalances with flexible employees. Employees must be able to perform upstream and downstream operations to make the pull process effective. After the employee meets the basic requirement of one up, one down, management should provide incentives to increase his or her flexibility. A cap may limit the maximum number of operations in which an employee can become certified. Because flexible employees are required to rotate frequently through certified operations, this cap is whatever reasonable number of operations an employee can be expected to perform over a given period of time. After employees have reached the certification cap, they may increase their pay-for-skill points by attaining certification in more difficult tasks (while forfeiting certification in simpler tasks), or becoming a team leader (see Figure 10.4).

Team Incentives

A TQC process team should produce total quality units in a linear fashion. Each process team's linearity index of TQC units is measured against the daily rate. Team passes (or two operational passes) of nonquality units are subtracted (as is an overproduction or underproduction) from the daily rate. Each process has a linearity goal. Employees have team incentives for attaining that goal. As the process matures, the goal of TQC units approaches the daily rate. After attaining 90 percent linearity against the daily rate of TQC units, the team is eligible for higher team incentives. The more "support" people on the team incentive plan (such as engineering and quality staff) the better. This fosters a team atmosphere and envelops the support team in the production-process focus.

Employee Involvement

The final element of the pay program is participation in the employee-involvement program. This differs from the traditional "suggestion" program in several respects. First, the employee-involvement program includes a formal response process. Employees make process-improvement suggestions regarding the SOE's (sequence of events) non-value-added categories. If employees feel valued and see their suggestions implemented, they make valuable suggestions more frequently.

The power in any production line is the employees—their desire and their knowledge. Often, suggestion programs flounder because resources are not available to respond to suggestions.

Suggested improvements typically have the biggest impact on design and manufacturing engineering. Suggestions should go above the "identify-problem, gather-data, isolate-root-causes, and monitor-solution" steps. Teams, not individuals, should make suggestions. Teams should also be given the incentive to make suggestions and not necessarily to find solutions. Also, the more "support" functions imbedded in the team, the greater the success.

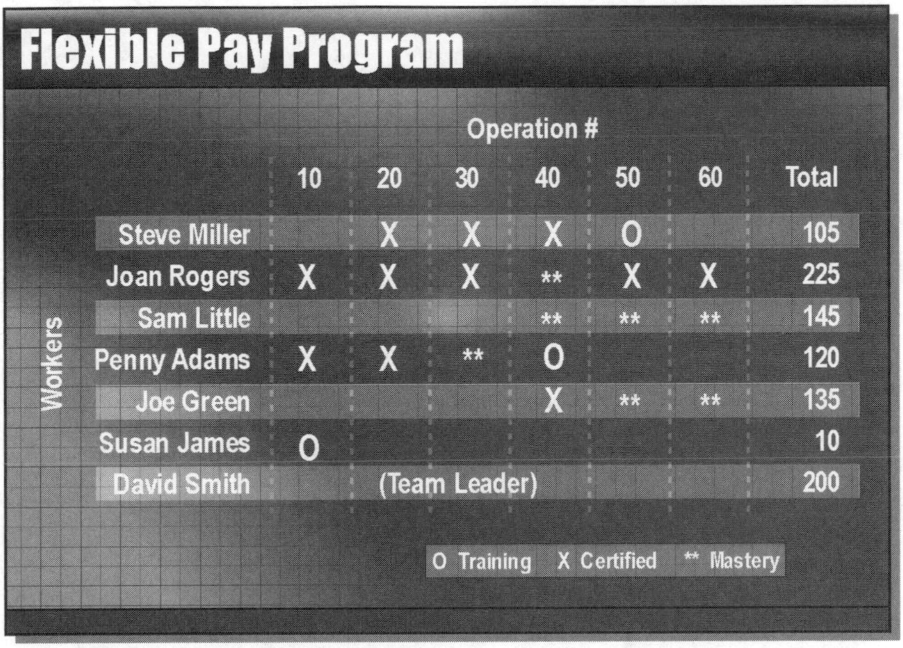

Flexible Pay Program

Workers	Operation #						Total
	10	20	30	40	50	60	
Steve Miller		X	X	X	O		105
Joan Rogers	X	X	X	**	X	X	225
Sam Little				**	**	**	145
Penny Adams	X	X	**	O			120
Joe Green				X	**	**	135
Susan James	O						10
David Smith	(Team Leader)						200

O Training   X Certified   ** Mastery

Figure 10.4 — Flexible Pay Program

Typical Demand Flow Manufacturing Pay Program

Job: Production Assembly Employee

Certification-Level Job Band: 4

Job-Band Rate: $2,000 per month

Market Value: $2,250 per month

Flexibility Factor: Pay-for-skill-point ranges described in Figure 10.5

Figure 10.5 — Pay for Skill Ranges

0 - 100 points.	Base pay.
101 - 150 points.	Base plus 10%.
151 - 200 points.	Base plus 20%.
201 - 250 points.	Base plus 30%.
251 - 300 points.	Base plus 40% (cap).

Team incentive: 1% for 75-80% linearity;
2% for 81-90% linearity;
3% for 91-95% linearity;
4% for 95-98% linearity;
5% for linearity above 98%.

Organizational incentive:
6% of base at risk;
6% lost at <80%;
6% paid based on 81-100% attainment;
over 100%, 50/50 distribution of found money.

Consider the following example. Cheryl Baker is a TQC employee. She has attained 180 flexibility points. Her process has averaged 84 percent linearity over the last three months. During the past year, the organization/division has exceeded its profitability goals and has a found a money "pot" of $200,000. Some 200 people are in this organization. Cheryl is paid as follows:

$2,000 base, plus $400 additional flex pay each month

Team incentive: $40 per week bonus for 84 percent linearity (2 percent of $2,000), paid quarterly, equals $520

Organizational incentive: 50 percent division of found money ($200,000), $100,000 in an employee pool, divided by 200 employees, equals $500 paid annually (this could be scaled based on earnings)

Employee involvement: Cheryl has made 46 process-improvement suggestions. Based on the total number of suggestions and the total savings received by the company, employee share of the savings is equal to $2,000, which is paid quarterly.

As compared with a traditional package of $25,000, Cheryl's total annual compensation package is $33,100, broken down as follows:

Base: $24,000

Flexibility: $4,800

Team incentive: $1,800

Organizational incentive: $500

Employee involvement: $2,000

Direct labor as portion of product cost: 5%

Savings to the company include the following: $100,000 found-money compensation; benefit dollars from the suggestion savings; intangibles for flexible-employee and linearity achievements. Administration of such a program is not a minor detail, but consolidating payments and ranging flex values helps minimize the challenges. Was it worth the trouble? Yes, by many tens of thousands of dollars and greater employee satisfaction.

Certification and Mastery Criteria

Certification criteria for an operation must be clearly defined. These criteria must include technical work content, educational requirements, quality expectations, and the maximum amount of time that can pass between assignments at that operation. An "operation" is a combination of events grouped together based on the targeted cycle time of the flow process. An operation is not each and every sequence of events in a process but a grouping of the sequences. One-up, one-down operations may qualify as a "job" for job banding.

A team of production, quality, engineering, and human-resource employees performs the "job banding" function and creates the certification criteria one time. Some "jobs" or operational criteria may change over time, and a system should exist for modifying the bands or criteria.

Mastery criteria differ from certification criteria in two ways: the production of high-quality parts for a proven period of time, and the ability to train and certify others. Some employees may be skilled in a particular craft but couldn't train fish to swim. They stay at the certification level. For those who wish to attain mastery level at an operation, the company must provide an adequate training program—to "train the trainers."

Sample Certification Criteria

Job band: #14

Job description: PCB assembly

Process/line: Through-hole assembly process

Job content: Assemble through-hole components into PCBs. Verify previous operator's work content. Check TQC points as indicated on TQC operational sheet. Ability to operate "contact system" or "manusert" machines. Ability to solder components onto board after insertion.

One-up/one-down operations: Wave solder operation

Minimum rotation cycle: Touchup operation once every four days

Job value:

>Training: 10

>Certification: 30

>Mastery: 40

Quality requirements: Maintain process capability of less than 50 parts per million. Maintain operational Pareto of position. Two operational passes of less than one per week. Contributed first-pass yield at an in-circuit test operation of 99 percent or above.

With the clear definition of certification criteria, all employees in a process should have access to the obtainment criteria. Training programs should enable employees to reach higher levels of flexibility. After-hours classes can provide employees further opportunity to increase their flexibility, particularly of the vertical variety. Well-defined criteria and the corresponding training enable today's production employees to learn through their own initiative and test tech skills that can someday increase their value to the process as well as increase their pay.

Employees may lose certification in a process due to a failure to meet specified criteria (most likely insufficient work in position to maintain certification). This decertification process is acceptable to the employees if they feel they have control over where they work. If decertification occurs when an employee has no opportunities to exercise flexibility, the employee feels that it is an unfair action. The determination of who works where and when is a duty of the line's production supervisor. All supervisors need to be aware of minimum certification requirements and rotation intervals.

Seniority, Unions, and Training

Demand Flow Technology does not affect or attempt to change the seniority position established by employees before conversion. After the DFT conversion, the senior employee retains the compensation achieved to that point. However, no employee rides to the top on longevity alone. If senior employees do not cross train and achieve mastery, their pay will not increase.

Several unions have embraced DFT for the same reason that employees like it: it gives employees the opportunity to control their own destiny.

Under DFT, the average annual training expenditure per manufacturing employee for the world-class manufacturer may increase. New hires in DFT operations are screened on the basis of their adaptability to the flexible TQC process.

Marginal employees carried over from traditional manufacturing may feel pressure to achieve a level of quality and flexibility proficiency not previously required of them.

Three-Phase Training

An active training program is essential to enable employees to attain and maintain flexibility. The first phase consists of the non-technical training required for working in the process. The quality department provides Pareto and process-control training. The human-resources group provides team, employee-involvement, effective-meeting, and interpersonal-skills training. Employees receive information on the company, its products, customers, values, and mission. Ongoing training is available to all employees. It seeks to eliminate one of the short-comings of traditional manufacturing—where employees who help assemble a product have no idea what the finished product looks like and are unfamiliar with the company's goals. This inte-grative effort is a part of the team-building process and is, therefore, vitally important. For the new employee or employees undergoing DFT conversion, 40 to 80 hours of this type of training occurs over the initial three to six months. Every quarter, upper management should give "state of the company" presentations to all employees in an open and frank manner with adequate time for questions and feedback.

The second phase of training is offline, where employees learn basic production skills in a simulated production environment. Verification and TQC recognition from TQC operational sheets are also taught. Employees usually undergo three days to two weeks of offline training.

Online Training Emphasized

The third phase of training is online training of a trainee by an experi-enced employee. The trainee works in the process with an employee who has obtained a mastery level. The trainee may actually perform the work under the guidance and scrutiny of the skilled employee. The employee at the mastery level is responsible for the quality of the work being per-formed. During this phase, the trainee becomes familiar with the certifica-tion criteria for the operation.

These certification criteria include the technical aspects of the operation, quality criteria, and the frequency with which an employee must perform the function to achieve and maintain certification levels.

Management Structure Changes

All employees become members of a flexible multitask process that emphasizes improved quality and problem solving. The management and employees in world-class companies evolve from traditional toward "top down" cultures toward more participatory work environments. Roles blend as the mix of groups change. In a Demand Flow manufacturing structure, management and supervisory responsibilities integrate to include the following:

- Demand-based planning
- Purchasing
- Production
- Manufacturing engineering
- Design engineering
- Quality
- Accounting/finance
- DFT manufacturing systems
- Maintenance

Team Leader

Team leadership of the production environment comes in two forms. The first is the more traditional supervisory leader who has responsibility for the flow line or multiple flow lines. This individual has transitioned from the traditional role to that of a true team leader who encourages participation from the operators on the flow lines. The second type of leader is the operational team leader on each flow line.

These team leaders are usually skilled in all processes on the line and can work in the process when and where needed. They focus on three main performance areas and represent the operators in the achievement of the production of TQC units, the achievement of the daily rate, and the suggestions for continuous process-improvement ideas from the team. Both production and operational team leaders confer with other team leaders. They jointly monitor rate, quality, and training; mobilize support functions; and generally manage the process.

If a particular need emerges (for example, in design engineering), the team leaders address it. Operational team leaders usually rotate with other production employees. The production team leader has many responsibilities, similar to the production superintendent in traditional manufacturing. The production super-intendent's position gives way to the team management of flow manufacturing. Layers of management and interdepartmental politics decrease significantly from traditional manufacturing.

New Management/Employee Participation

Across the globe, management styles are evolving from traditional to participatory, with a few companies even pioneering a form of "socio production." The traditional management method in place for most of the 20th century was completely inflexible: people were told what to do and how to do it. More and more companies are moving toward a participatory style that involves job sharing, cross training, and flexibility. Employees are told what needs to be done, and the team figures out how to do it. In socio production, the team decides what to do and how to do it. Socio production is extreme and rarely seen.

Tradition Yields to Participation

On a scale of one to five, with one being least progressive and five being the most progressive, traditional management would be one, participatory would be three, and socio production would be five for the following management factors: task assignments, operation design, performance maintenance, organizational design, decision making, staffing, leadership, compensation, financial controls, training, technology, and people.

The differences among the styles are great and many. In traditional management, people are accountable for their own work; in participatory, they are accountable for their own work and the work of the team; in socio production, they are comfortable with ambiguity and change. Some companies are inclined toward participatory management and even socio production in some areas, while remaining traditional in others.

Demand Flow Technology and Participatory Management

Some confusion usually ensues concerning the relationship between Demand Flow Technology and participatory management. Some executives believe that when they adopt participatory management, they have become a world-class Demand Flow manufacturer. This is not true. Participatory management is one element of world-class Demand Flow manufacturing. It is one of the techniques used to implement Demand Flow Technology. Companies managed in the participatory style are just that.

They are not DFT unless they have implemented all other Demand Flow manufacturing techniques as well. The DFT participatory organization, with its flexible employees and operational decision making, evolves to a flat management structure with teams empowered to make decisions.

Employees Are Key

Establishing flexible employees is essential in a DFT environment. Flexible employees are key to successfully attaining world-class Demand Flow manufacturing goals. The compensation systems change to reflect the requirements of Demand Flow Technology. Traditional "job description" classifications with fixed reward systems are outdated and offer little encouragement for individuals to achieve their personal best. Pay systems must change when the traditional manufacturing lines are redesigned for the flow process. In Demand Flow manufacturing, flexible employees are vital to the success of the TQC production process.

Chapter 11
Total Quality Control

Demand Flow manufacturing's emphasis on total-quality focus is unbending in its pursuit of manufacturing excellence. Customer demand leads DFT companies to place a higher priority on quality than on production. DFT's enormous savings in scrap and reworked materials and lower overhead costs make possible this intense quality focus. Companies cannot achieve Demand Flow manufacturing without TQC, and they cannot implement TQC without an effective employee involvement program and a consistent flow process. Who knows the real work content of an operation better than the person performing the work every day? In a world-class Demand Flow process, employees are stimulated and offered incentives to strive toward a goal of total quality and continuous process improvement. An effective employee-involvement program must be responsive to be effective. While it may be impossible to implement all employee suggestions immediately, and while many suggestions may not initially appear to be "mission critical," employees must feel that they are valued members of the team and that their suggestions are receiving due consideration. The greatest cause of failure in employee-involvement programs is the unresponsiveness of management and support teams.

TQC in the Process

TQC is based on internal process control at the source of the work. A program of enforced problem resolution uses all employees, including finance and sales in addition to materials, engineering, production, and quality. All employees in the company are assigned to an employee-involvement team. They may not regularly attend meetings, but they are on the team and eventually play a role in improving the process. People from the same support department are assigned to different quality teams. Any individual or organization may be assigned an action item from the employee-involvement process. The goal is to eliminate organizational partitions and boundaries that limit process improvement. Management participation is mandatory.

A Streamlined Approach

Teams meet only for specific purposes. They pursue specific opportunities or resolve specific problems. They monitor the process, identify problems, isolate root causes, and eliminate those causes. The team shares the goals of process perfection, quality improvement, and cost reduction. The world-class company focuses on high quality, process efficiency, eliminating non-value-added work to lower production cost, total employee involvement, and increased customer responsiveness and satisfaction. The DFT company accomplishes these goals using a common technology that makes the best use of the company's people. TQC identifies customer expectations, establishes target values, defines process sequence and elements, determines process capabilities, and reduces process variables.

In-Process Quality

Demand Flow manufacturing's methodology of total quality control eliminates non-quality parts, processes, and products. This quality focus is the natural outcome of the initial design of the TQC operations and the involvement of people working on quality in the process. In traditional schedulized manufacturing, inspectors perform only a few nonproductive quality determination actions. In DFT, many quality checks and verifications occur throughout the process.

Quality is embedded into the process rather than checked outside of the process. The first thing a production employee does when pulling a part or product is to validate the previous work content through a TQC. Even the best employees can and do make mistakes. If the operator finds an error during TQC, he or she passes the work back to the operation where the error originated.

Any "two operational" passes are forbidden. The two-operational pass occurs when one team member makes an error and another team member doing TQC does not identify it. Also known as "team" passes, two-operational passes do not count toward linearity-attainment goals. Quality teams identify quality problems and request their elimination. Management must listen closely to operational employees about design flaws that contribute to quality problems.

TQC Operational Sheets

Illustrated TQC operational sheets not only show the operator the work to be performed, they also show the work to be verified and the required total-quality-control inspections (see Figure 11.1). Employees can verify work performed at the same operation, but they can never do a TQC inspection of work performed at the same operation. Any time multiple ways to perform work exist, but only one way is correct, the work requires a TQC at another operation. Total quality does not require that employees verify everything, only very specific elements of work. Total-quality methods precisely direct the employee to specific points of verification and total-quality validation. These TQC points are also guides to design engineering for "designed for defect" points, which must be corrected. Sometimes, TQC points are covered during the process, preventing subsequent checks. These products are identified as acceptable quality level (AQL) products until the design defect is eliminated and a feasible TQC design is incorporated. Other points, while verifiable, are not "failsafe," such as when multiple ways to perform the work exist and each way is correct. The employee-involvement process must identify these design dilemmas and prioritize their elimination. The transfer team reviews new products released from design to eliminate designed-for-defect points.

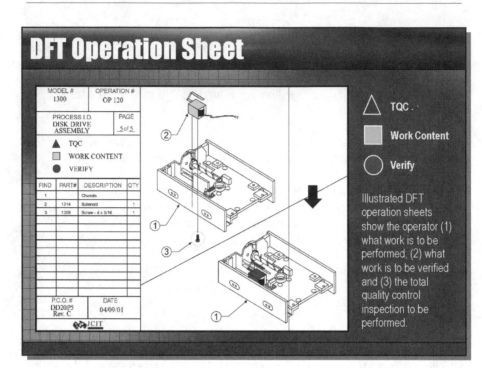

Figure 11.1 — DFT Operation Sheet

Targeting Process Capability

Quality products are the direct result of a quality process. In total-quality Demand Flow manufacturing, designing a quality process precedes the manufacturing of products. Total quality control at the source eliminates the potential for quality problems and breakdowns. World-class manufacturing designs the product for total quality, designs the process to create a quality product, and then assigns the responsibility for total quality to the employee. The focus on quality continues with process-capability (Cp) techniques, which are defined as the ratio between the design specification width and the observed process capability. To calculate the value for Cp, divide the design specification width by the observed process specification width. The quality process targets the design center, or the nominal value (see Figure 11.2).

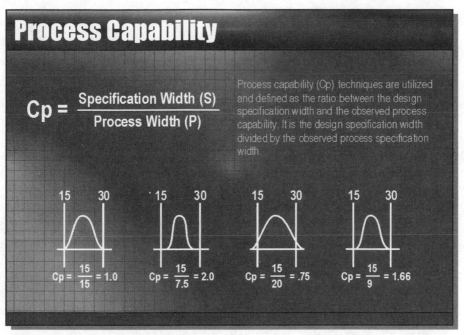

Figure 11.2 — Process Capability

Total quality control places a high priority on identifying and eliminating process variations. Variations create a broad process specification and result in inconsistency and poor quality performance. Process capabilities must fall within design specifications, and the target value must equal the design center. As the process capability value increases from 1.00 to 1.66, on to 2.00 and beyond, in-process inventories and scrap decrease. To ensure adherence to the targeted design center, measure the process capability (Cpk) related to the design center.

$$CP = \frac{\text{Specification Width (S)}}{\text{Process Width (P)}}$$

$$Cpk = (1 - [K])\, Cp$$

$$K = \frac{D - (X)}{S/2}$$

D = Design Center

(X) = Process Average

The total-quality Demand Flow process always assumes actual performance will fall within the process specification of a part or product. This means processes drifting out of control are detected prior to falling outside the design specification and creating scrap and rework.
A production process with a process capability (Cpk) greater than one but less than the designed Cpk, is producing good products in the eyes of the customer and the design specification—but it is also a warning sign that the process may be drifting out of control. The objective of the TQC process-perfection program is to tighten the observed process capability while the design specification remains constant. This causes the Cpk ratio to rise. In the TQC Demand Flow process, if the process capability is less than one, the company must employ the dual-card kanban technique and in-process inspection. When Cpk is less than one, this indicates that the process cannot consistently meet the design specifications. The dual-card kanban technique enables production to start in a quantity greater than the consuming process's pull quantity. The additional production quantity is necessary because of yield and/or process problems. Until it resolves these problems, the company makes a temporary retreat and produces larger quantities of units each time. On machines with a Cpk of less than one, scrap or rework occurs, and more units must be produced to compensate for the yield problem.

Experimentation of Design Techniques

Design experiments are an essential tool for identifying and eliminating dangerous variation for the purpose of establishing and maintaining control over any given process. A process in statistical control is repeatable, stable, and predictable. The total-quality manufacturer is interested primarily in prevention techniques. The data gathered is simple, visible, and specific to the operations and machines. Quality products are the direct result of a quality process. The manufacturer should focus on fixing the process rather than the more costly alternative of inspecting and fixing the product. Quality must be inherent in the product and process and not the result of external inspection. Every defect has a cause; every cause, a solution. The solution is found through the TQC/employee-involvement process.

Process Perfection Program

Every operation in the process should maintain an operational Pareto chart to track opportunities for improvement and the number of such opportunities. The production employees maintain this information in quality notebooks. Each employee, each team, each support organization, and the division as a whole have prominently posted Pareto charts identifying opportunities for improvement (see Figure 11.3). The Pareto chart highlights any dangerous "designed for defect" flaws and other known problems so the employee can easily track occurrences. The highest priority on the Pareto chart is identified from the left, which is typically the problem/opportunity with the greatest number of occurrences. The Pareto chart also continues the trend of meeting product specifications within the process specification. Teams identify problems and opportunities and assign them to the related areas or support teams.

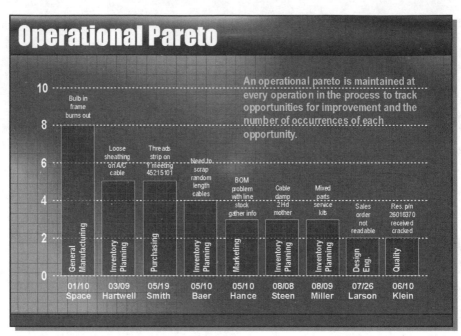

Figure 11.3 — Operational Pareto

Problems Are Opportunities

DFT companies treat problems as opportunities, and they prominently post this information. Root causes are identified, responsibility assigned, and data gathered. Control charts ensure process capability. Problems are identified prior to the process exceeding design specifications. The process may drift out of control well before a "bad" part or product emerges. Process-control charts identify problems/opportunities so that appropriate teams and individuals can take action. Experimentation of design techniques also improves supplier tracking on problem parts. Manufacturers often discover that they have attempted to drive a particular part well beyond its capability or beyond the capability of the product/process. The operational Pareto chart is the format for collecting data to identify this situation and point the way to the solution. A "fishbone" technique identifies all possible causes of a problem and isolates the probable cause via cause-and-effect analysis (see Figure 11.4).

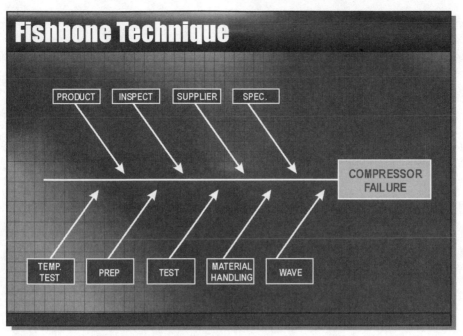

Figure 11.4 — Fishbone Technique

Continual-Improvement Goal

Everyone on the process-perfection and employee-involvement team must be trained to brainstorm and maintain Pareto and control charts, and understand fishbone analyses. The TQC Demand Flow manufacturer must institute a pervasive quality focus at both the employee and machine levels. For every problem, there is an opportunity. For every opportunity, there is a plan. The problem-solving process has its own audit trail: specifically, who has the problem, how long have they had it, and what are they doing about it. The entire process is highly visible. Permanent solutions replace the original causes, whether the solution involves fixing the manufacturing process, redesigning the product or process, or training employees. The process undergoes continuous monitoring. If the targeted problem goes away, the true cause has been found, the problem is solved, and the company moves on to the next opportunity. The goal is continual process improvement. The standard is perfection against process capability at the point of value.

Excellence Is the Focus of All Personnel

DFT focuses relentlessly on process perfection and problem solving. The ongoing effort includes everyone in the company. Problem solving at the world-class company is everyone's responsibility, from the entry-level employee to the chief executive officer. Everyone is associated with a TQC team. After all, employee involvement is the most powerful aspect of the program. The resulting quality improvements mean higher customer satisfaction, improved productivity, and improved profit.

Chapter 12
Demand Flow

Making the financial and management reporting systems consistent with the responsive Demand Flow business strategy is a prerequisite to evolving beyond a departmental financial-costing system. With flexible processes and production employees, tracking direct labor is not a priority. Also, direct labor plays no part in the application of overhead costs. Demand Flow's focus is on material and overhead costs, *not* direct labor. It is very confusing to implement Demand Flow manufacturing without a corresponding change to the financial system. Labor-based costing techniques can create an illusion of negative growth as the highly productive flow process shrinks the labor basis.

Financial control and management in a Demand Flow manufacturing environment are just as important as they are in traditional manufacturing—but also significantly different. Traditionally, companies have used the old "three-bucket" approach:

- Raw material in the storeroom (RAW)
- Work in process (WIP)
- Finished-goods inventory (FGI)

In schedulized manufacturing, the company schedules individual work orders and issues material to build a specific quantity of a specific assembly or fabricated part. Each unique work order is carefully tracked in detail to collect the costs related to issued material, production labor, and corresponding overhead. No reporting of WIP variances occurs until the production work order closes. Physical inventories have proved that the actual WIP inventory seldom matches the book WIP inventory. Traditionally, it takes weeks to complete the production assigned to a work order and for the product to go through the process—and to report the corresponding results.

Accuracy, Availability, Adaptability

The finance department's primary objective in Demand Flow manufacturing is to provide management with accurate and timely information. Financial organizations must be adaptable enough to produce the precise information required by DFT management. In addition to accuracy and timeliness, the costs of collecting and analyzing data must be reasonable. The DFT manufacturer must look at the product as a flow process. Data collection should wait until a product is complete and backflushed. The product is built in the flow process, and the data is collected at the end of the process. Relevant data from the financial accounting system should include:

- Cost of the product—the correct product pricing
- Value of on-hand inventory—for internal and external reporting purposes
- Cost controls—to monitor the cost of the production and operating activities
- Simulation—the ability to adjust product volume and mix to evaluate various margin scenarios

While the product is in process, the manufacturer should collect only the data on scrap. Scrap data must be collected whenever and wherever scrap occurs. Otherwise, the bill-of-material data is the only guide for material to be purchased and consumed in production. This planned consumption occurs when the product leaves production and its parts are backflushed (relieved from raw and in-process inventory).

Monitoring Product Costs

Product-cost accounting in the Demand Flow manufacturing environment includes the planned material consumption from the bill of material measured only at the final backflush point. Material overhead (acquisition cost) applies as a percentage against the product's raw material at its standard material cost, and any extra usage and scrap transactions are recorded by exception. Company-wide homogeneous overhead applies against the total product cycle time. Under the Demand Flow business strategy, FlowBased™ costing techniques establish a standard product cost by taking the material content of each product and adding the accumulated operational costs (overhead). Actual product cost per unit is calculated by totaling operational costs for the process during a given period and dividing that amount by the number of units produced during that time period. This calculation yields actual operational cost per unit.

Labor De-emphasized

Demand Flow companies have demonstrated an uncanny ability to focus on material cost—to a far greater extent than most schedulized companies. For years, companies in the United States have emphasized labor costs, even though the labor costs on most efficient processes have dropped to between 5 and 15 percent of product cost. Meanwhile, material and over-head costs have soared to make up 85 to 95 percent (see Figure 12.1) of total product costs.

The dilemma the typical, traditional schedulized company faces is whether to shut down its production line, even though lack of demand clearly indicates that a shutdown is the right decision. For years management has worked from the assumption that an idle line would cost the company $1,000 per minute. Unfortunately, management keeps the lines running and manufactures excess inventories that sell at a loss. Under the Demand Flow business strategy, the line volume adjusts to match market demand. Demand Flow manufacturing focuses on the lion's share of product cost: material and overhead. The basic premise in DFT cost accounting is that the direct labor portion of the product cost is shrinking and is too small to warrant detailed tracking. The total product cycle time is short enough that the labor portion of inventory value moves through the product process quickly. The amount of inventory in the Demand Flow process is relatively small.

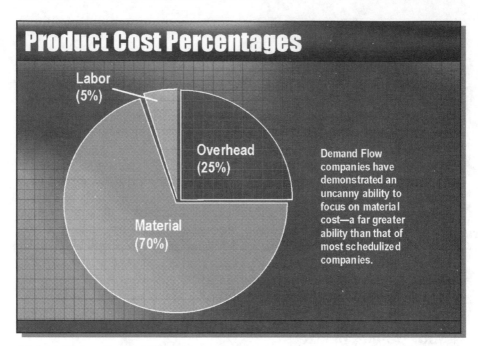

Figure 12.1 — Product Cost Percentages

Flow-Based Overheads

Demand Flow manufacturing applies homogeneous overhead to total product cycle time for several reasons. Applying homogeneous overhead to total product cycle time does not penalize production for becoming more efficient. Total product cycle time has a direct relationship with the amount of homogeneous overhead consumed and the rate of inventory turns. Traditionally, applying overhead to direct labor increases the amount of overhead per production headcount if production becomes more efficient and labor work at an operation is eliminated.

Applying overhead to material dangerously increases the amount of overhead applied, especially when an assembly or fabricated part is subcontracted and the standard material cost is increased, or if two products with similar work content are made from materials of vastly different value.

Applying overhead to material diverts the company's focus toward reducing material costs and inventory turns—essentially creating a supplier cost-reduction program. This also diverts attention from the production process, which in turn jeopardizes the benefits of the continuously improving TQC flow process.

Establishing the Standard Product Cost

The standard product cost still includes the following:

- Material
- Labor
- Overhead

Labor (direct and indirect) is not relevant to an operational-efficiency standard. Because production employees flex (or move) to fill "holes" created in processes that are producing at less than capacity, knowing exactly which person produced what quantity is meaningless information. Labor costs become an element of overhead costs.

Overhead costs now contain common labor costs, and this homogeneous overhead cost applies to the total product cycle time for each product. Other variable overhead costs can also be charged per square foot of manufacturing space occupied, per hour of planned usage, and product-specific resource requirements.

The total annual overhead cost is the sum of the common labor costs, at an average hourly wage (labor/hour), plus the fixed overhead costs per year. The sum of the costs to convert purchased material into a finished product includes fixed costs, variable costs, and labor costs—forming a homogeneous overhead pool. This homogeneous overhead pool contains conversion costs that apply to all products. Also, a variable overhead cost may help account for extraordinary conversion costs resulting from the use of special machines or resources.

Only products that require these costly resources absorb these extraordinary overhead costs. This extraordinary overhead cost could also be allocated based on the square footage the manufacturing process occupies (see Figure 12.2).

Figure 12.2 — Calculating overhead costs in a Demand Flow manufacturing environment

Homogenous overhead ($AOH) = (P x $AVG) + $OH

P = Planned annual production employee hours (the sum of planned annual volume x SOE labor times)

$AVG = Average production wage per hour

$OH = Overhead costs per year (fixed and variable, but not extraordinary conversion costs)

The basis for the application of the homogeneous overhead is the total product cycle time of each product. To achieve this, a cost of homogeneous overhead per TP c/t hour ($TP c/t hour) must be established where the following applies (see Figure 12.3).

Figure 12.3 — Calculating cost of homogeneous overhead per TP c/t hour

$AOH = Annual homogeneous overhead

Annual planned TP c/t hours = annual planned volume of each product times its TP c/t, in hours

Example:

	Planned Volume	Total Product-Cycle Time	Annual Planned TP c/t Hours
Product "A"	1,250 units	6.8 hours	8,500 hours
Product "B"	568 units	6.8 hours	3,862 hours
Product "C"	1,100 units	3.0 hours	3,300 hours

Annual planned TP c/t hours = 15,662

To calculate $TP c/t hour:

$$\frac{\text{Homogeneous overhead per TP c/t Hour}}{(\$TPc/t \text{ Hour})} = \frac{\$12,000,000}{15,662}$$

$TP c/t hour = $766 per TP c/t hour

Applying homogeneous overhead per TP c/t hour on each product's total product cycle time may not account for all the costs of producing the product or product family. Two products could have the same TP c/t, but one of the products may require the use of an expensive machine, while the other does not. It may be more equitable to apply the cost of the expensive machine only on the product that requires this resource. If this is the case, the extraordinary cost of conversion should be applied to that specific product or family of products.

For example, assume the following:

Product "A" requires the use of a $500,000 machine

Product "B" does not require this machine

Product "C" does not require this machine

The cost of this machine should be applied only to Product "A." To allocate this extraordinary overhead cost, the square footage of the process could be used as a basis for the allocation, and only Product "A" bears the cost. Based on this extraordinary overhead cost, a cost per 1,000 square feet of manufacturing occupancy that is applicable only to Product "A" can now be computed.

The application of the extraordinary overhead costs such as depreciation, specific machine maintenance, and utilities is accomplished by structuring specific accounts/sub accounts into a cost pool that comprises machine hours related to the specific activity that consumes the specific machine process.

Caution: Many companies tend to over manage extraordinary overhead costs. Do not microscopically explore your facility for these costs. They should be readily apparent.

Purchased Material Content

To establish the total product cost, the purchased material content also comes into play. The bill of material in DFT must be 100 percent accurate to backflush inventory. Each product's bill of material is costed out at a total raw-material standard cost. Purchase price variance results from comparing the actual purchased price versus the raw-material standard cost. The cost of carrying inventory should also be added to the total cost of the raw material, not to the standard raw-material cost.

Figure 12.4 — Calculating the total cost of raw material

RM standard cost

+   Acquisition cost

Total RM standard cost

To establish product cost in Demand Flow cost accounting, the factors in Figure 12.5 apply:

Figure 12.5 — Establishing standard product cost for Demand Flow manufacturing

Equation:

$$\begin{array}{r} \text{\$TP c/t hour} \\ \text{x TP c/t} \quad \text{extraordinary variable overhead costs} \\ + \quad \text{Total RM Standard Cost} \\ \hline \text{Product Cost} \end{array}$$

Example:

MODEL 494X
(Product "A")

Total Product Cycle Time	=	6.8 hours
Total RM Standard Cost	=	$8,371.50
Homogeneous Overhead per TP c/t hour	=	$766
Extraordinary variable overhead costs	=	$65 per unit (allocation of cost of specific expensive machine)

The resulting product standard cost would be:
```
  $  5,208.80  ($66 x 6.8 TP c/t hours)
+ $      65.00
+ $  8,371.50
  $ 13,645.30
```

In the analysis of actual costs versus planned standard costs, the actual overhead and actual TP c/t are used. Physical audits of TP c/t to monitor progress and to prove the computerized calculation of TP c/t are common responsibilities of DFT financial organizations. In actual product cost analysis, any actual material scrap costs are added to the actual bill-of-material costs. Although very difficult to accomplish, an actual overhead variance is expected and achieved by reducing the total product cycle time. Management must place a high priority on reducing this variance and budget accordingly.

Valuing In-Process Inventory

In trying to obtain the value of RIP (raw and in-process), companies must remember that in a high-inventory-turn Demand Flow process, the added dollar value of in-process inventory may be insignificant when it is contained within a single flow process without staging points. The outward flow of RIP material must be measured at the end of the process. Quantities in the process can be determined through understanding the process capacity and whether the process is full or empty.

In calculating overhead value in RIP inventory, we consider that at some early stage of the flow process, the product being produced is 10 percent complete. In the middle of the flow process, it is 50 percent complete. Toward the end of the flow process, the product is 90 percent complete.

In a takt-time-designed Demand Flow line, the average overhead cost per unit in process is calculated by dividing the standard overhead cost per unit by two. This homogeneous principle assumes that each unit in process is, on average, 50 percent complete. The overhead value in RIP is determined by multiplying the sum of the number of operations plus the in-process kanban quantity times the standard overhead cost divided by two. For example, see Figure 12.6.

Figure 12.6 — Overhead value in RIP

Assume the number of in-process kanban is 12 and the process includes 12 different operations. Given these parameters, the overhead cost per unit is $766, and the line is a balanced Demand Flow process.

The calculations are as follows:

$$\frac{766}{2} \times 12 = \$4,596$$

The value of RIP is:

Material value = $ 1,260,000

Overhead value = $ 4,596

Total RIP value = $ 1,264,596

Changing Departmental to Flow-Based Costing

In changing from departmental, schedulized manufacturing with labor-based cost accounting to flow-based costing techniques, the production process must change first. After a company changes to a DFT production environment, the financial organization must audit the flow process and calculate costs. The role of accounting changes from labor and work-order tracking in multiple departments to flow-based process accounting. Financial managers become more influential through their decision making and analysis, in contrast to their traditional role of tracking transactions. Once the company has embraced the Demand Flow business strategy, the emphasis shifts to quicker customer response (no late shipments), zero working capital, increased inventory turns, and eliminating unnecessary overhead. Financial managers must understand Demand Flow business strategy, as they are expected to help lead the transformation toward an elite zero-working-capital company.

Chapter 13
Achieving a Commitment
to World-Class Excellence

Adopting Demand Flow manufacturing technology is not an optional course for a company committed to becoming a world-class manufacturer. Demand Flow is a technology that, once adopted, must cross all organizational boundaries and receive management's full support. It is a company-wide program in which management must seek out and win over any pockets of resistance.

In the 1990s, many companies came to understand that they had to make the change to flow-manufacturing mentality. Faster response to customer demand and increased speed in all management activities characterize the Demand Flow business strategy. Adopting this strategy provides significant benefits that are not obtainable by using tools and methods of the past. The Demand Flow manufacturing methodology focuses on the two major elements of product costs: material and overhead. Demand Flow manufacturing techniques create powerful production process that rely on pull systems, with in-process quality as the primary objective. Achieving the elite goal of world-class manufacturing requires the pursuit of nontraditional goals and the implementation of nontraditional process-management methods.

Several different tools assist in managing the Demand Flow process, including the following: Total product cycle time (TP c/t), flow rate, linearity-index measurement, team passes for non-quality items, the number of line stops or time per problem, in-process kanbans, inventory turns, and employee involvement.

Management must be committed to and supportive of Demand Flow manufacturing technology. The path is difficult, complicated by mindsets that believe "we're unique" and "it can't work here"; but the tremendous benefits are essential for significant growth and to meet the financial objectives of the corporation. Management must stay committed and supportive and continually express two clear messages:

- All employees are critical to the success of the company, and each has a voice in the company's business strategy.

- There is no question of management's commitment, and the company's strategic direction is focused on flow manufacturing.

Measuring a TQC Demand Flow Line

The most important goal of a world-class production process is total-quality products. Quality is never compromised for any reason. After locking down quality, the company's next goal is to make quality products equal to the daily rate. If the daily rate is 100 units, the goal is to make 100 quality units—not 95 or 105. The method of measuring and auditing a Demand Flow process is significantly different from measure-ments in traditional manufacturing. With the flexible employee, measuring individual perform-ance is not possible. All measures are team measures.

Cycle Time Monitored

The Demand Flow manufacturer monitors the TP c/t of the flow process. If the calculated TP c/t is one hour, the manufacturer physically reviews the production process and audits this time. Unfortunately, it is not possible to audit actual TP c/t remotely. An audit must be physical.

If the company has made process improvements since the last audit, it may be reasonable to expect a reduction in the TP c/t. The manufacturer then "under-absorbs" overhead for that process. Because "holes" can occur in the DFT flow line, flexible employees using in-process kanban signals flex (move) across the operations of the line. As a result, the product being manufactured and undergoing a physical audit of TP c/t may periodically be in an operation where no actual work is occurring. This is because an operator has flexed to another operation for a short time. Thus, the audited TP c/t often includes some variance, but it must never exceed 120 percent of the calculated TP c/t value.

The linearity index also undergoes close scrutiny. After achieving a 92 to 96 percent linearity index against the daily rate, the Demand Flow manufacturer may start to measure actual production flow rates at the back of the process. To some companies, this range of 92 to 96 percent linearity may seem to be a modest target. This is not the case in the majority of flow lines. In fact, a line with an 85 to 90 percent linearity index against flow rates is an excellent flow line—on the brink of world-class performance. Remember that the linearity index requires the measurement of TQC units that are completed, versus the daily rate required, with the target to produce the exact daily rate, no more and no less. Many companies struggle initially to meet this criterion.

Support teams should be close to the process they support. The production team leader is responsible for marshaling these resources when process problems occur. Support teams cannot be stationed remotely. They must be in a position to respond quickly to emerging process problems.

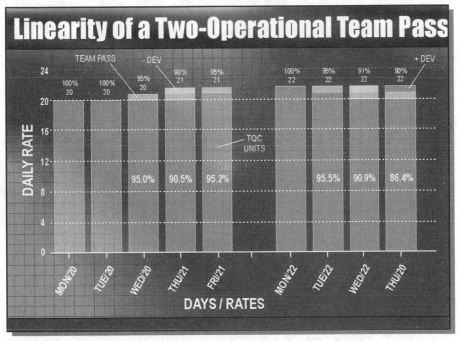

Figure 13.1 — Linearity of a Two-Operational Team Pass

Team Pass

A team pass (or two-operational pass) is another important measure of total-quality flow-line performance. The team pass occurs if the unit was produced incorrectly and validated incorrectly at the following TQC operation. The unit is returned for rework. Although the reworked unit is now perfectly acceptable from a customer, sales, and financial perspective, it does not count toward the daily production linearity goal. A team pass can invoke powerful peer pressure in the process. If a unit requires rework, it is tagged with the pass of the team responsible for the rework. The unit goes through the entire balance of the process with the team pass.

The product is not credited to the team goal, and the team cannot make up for the team pass and regain credit for the reworked unit. Just like defective work in a customer's hands, non-TQC work in the plant represents a non-recoverable situation. The percent of team passes and deviations against the daily rate is tracked each day. See Figure 13.1 for the linearity results of a two-operational team pass.

Management Lights

Switches on the line, accessible to all employees, provide management with accurate and continuous information on how the Demand Flow line is running. Each Demand Flow line has one set of management lights to communicate the overall status of the entire line. This management light system has green, red, and yellow lights. Each production employee has immediate access to a red and yellow switch. It is not necessary to oversupply each operation on the line with switches, but switches and lights should be accessible to each process and defined in the product synchronization.

If no switches are activated on the line, the management light system indicates a green condition. Green exists automatically in the absence of another status condition. If an employee identifies a need to replenish a raw-material kanban, he activates a switch that flashes a yellow light at the process and on the management lights. The yellow light indicates to the material handlers roving the process that a material is in need of replenishment. They must go to that location, read the kanban identifying the part and its point of supply, and get material back to the line within a predetermined replenishment time period.

If for some reason operators on the line cannot work in their primary position or one up, one down (for example, during a machine breakdown), the status switch is activated to indicate a red-light state. The red light does not indicate that the entire process has stopped, only that a problem at a particular operation has the potential to shut down the process. In effect, the red light becomes a call for help from the line operators, indicating that an issue needs resolution as soon as possible. The light brings the appropriate support people to resolve the problem. Management's ideal light sequence—indicating minimal delays, adequate people, and sufficient material—is yellow, with intermittent green flashes and occasional reds.

Timely Replenishment

It is perfectly acceptable in a Demand Flow process for employees to replenish their own material. By definition, RIP areas are close to the related process. The need for operator replenishment is an operations-management decision and should take into consideration the process takt time. The shorter the takt time, the less time the operators have to leave the line. With longer takt times, operators have more time for material replenishment. If production employees are out of material, they move in the direction of the pull, which in this case is to the appropriate RIP to replenish material. This occurs on an exception basis: employees replenish their own material only when the material handler cannot do so in the allotted time. If employees are expected to replenish their own material on a regular basis, this activity should be included in the TQC sequence of events for that process.

Management's Accordion

The Demand Flow production process can produce between 100 percent and 50 percent of the designed demand-at-capacity volume, with as much ease as opening and closing an accordion. Management can pull half of a line's employees, and the line will function smoothly, but only at about half its normal rate. Management also can fill or create voids on the Demand Flow line to adjust the daily rate with the aid of flexible employees.

Reporting in a Demand Flow Process

Employee flexibility in a Demand Flow process means that tracking and reporting for individual employee performance is unfeasible and unnecessary. The reporting in a Demand Flow process is simple, direct, and meaningful. Labor and machine content per product/process in total hours per unit is known and reported, which determines how many people are needed in the process at a given rate. Cycle time—both total product and operational—is also monitored, as is each team pass. The number and duration of line stops are tracked. Inventory levels of purchased material and in-process kanbans are monitored and reported. Companies expect continual improvement in these measures, driven by the employee-involvement program. Pareto charts, control charts, and fishbones are utilized extensively. Reporting on team performance is highly visible. Managing a production process also involves maintaining employee certification charts and criteria, as well as determining on a daily basis where each employee's primary position will be.

Chapter 14
The Role of Information Systems in DFT

The technology and methodology of Demand Flow manufacturing change the role of the computer system. Many of the execution techniques of Demand Flow manufacturing don't require a supporting system. However, information-technology support can make it easier to do the following:

- Calculate operational cycle time
- Calculate total product cycle time (TP c/t)
- Establish a demand-based daily-rate production plan
- Manage kanban tools
- Administer demand-based forecasting of material from suppliers
- Calculate the application of overhead to TP c/t
- Calculate the resources (key machine utilization and the number of people) required to meet the daily rate
- Improve process efficiency for management reporting
- Assess production linearity for management reporting
- Compare actual versus planned production-employee efficiency
- Manage TQC operational sheets
- Maintain efficient supplier management

Demand Flow manufacturing requires substantially fewer system transactions, decreasing the company's reliance on information technology to manage transactions. However, Demand Flow manufacturing without a computer may in fact be easier than MRP II *with* a computer (see Figure 14.1).

Toyota and many other major flow-manufacturing companies do not use computers for the planning and execution of their flow-manufacturing processes. However, system support can assist in line design, daily management and planning, and in the forecasting and control of material movement to and from suppliers.

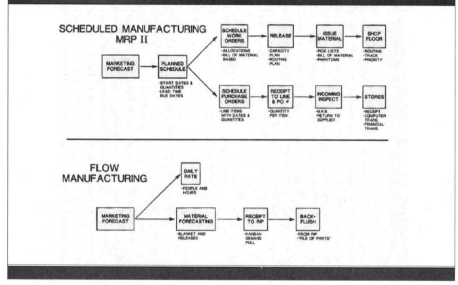

Figure 14.1 — Computer Role in Flow Manufacturing

Traditional System Evolution

Material planning initially meant exploding a bill of material solely for calculating the purchased material needed for a product. As the traditional computer system evolved into MRP I, and as the bill of material evolved into a multilevel subassembly bill of material, manufacturing software became more sophisticated. In addition to directing a company what to buy, MRP I also showed what to make and when to schedule the start of production.

MRP II brought additional enhancements in scheduling and continued the evolution of a sophisticated routing and tracking system. It tracked each assembly or fabricated part and the corresponding materials, labor, and overhead throughout the process. It became a scheduling tool, a management tool, a batch-producing economic order quantity (EOQ) production tool, and an operator-efficiency-measurement tool. Work orders for multi-level assemblies or fabricated parts were tracked through independent departments and work centers, along with the corresponding material and labor. The system also kept track of purchase orders and the related delivery dates and quantities. Additionally, it tracked each work order on the shop floor. It told planners what to schedule and storeroom employees what parts to issue. The system completely tracked the location of materials in the storeroom and in the WIP (work-in-process). MRP II evolved into sophisticated labor-based accounting and planning systems that enabled the tracking of multilevel products through multilevel departments, while offering some promise of controlling a complex manufacturing process.

Ruled the Roost

MRP II systems became big, bulky, and difficult to manage in real time. They tended to proliferate management reports—in effect, the MRP II system is management by report. The manufacturer's dilemmas center on which report to use and how to use it most effectively.

The MRP II systematization of the push philosophy encompasses a sophisticated system of the following: order tracking, order launching, expediting, queue reporting, multilevel bills of material, job costing, dispatch lists, shop floor control, expedite lists, capacity requirements planning, kit issues, work-order receipts, queues and lead time, work-order scheduling, labor-based accounting, and management by report.

Formal Path to Demand Flow Manufacturing

Although the Demand Flow manufacturing pull philosophy has drastically simplified requirements, information technology becomes a valuable tool when used for particular tasks. Many of these tasks are traditional in that they are very similar to the functionality previously provided in MRP and MRP II, but many tools are specific to flow requirements. The following lists the main areas where the systems support is most useful in Demand Flow manufacturing:

- Backflush transactions to remove material from the process
- Engineering operational evaluations
- Operational-line balancing
- Daily process-linearity calculations
- Linear-rate indexing
- Kanban management
- Kanban pull sequencing
- Kanban sizing
- Calculating operational cycle times
- Calculating total product cycle times
- Financial applications of overhead to TP c/t
- TQC operational-sheet design and management
- TQC sequence of events
- Developing demand-based planning (rather than scheduling)
- Flexible demand-based forecasting for suppliers
- Process accounting standards from the sequence of events

DFT emphasizes management by eyes and management (via a visual factory) through people, rather than managing by report. Demand Flow manufacturing techniques substantially reduce the number of reports, part numbers, and eventually even suppliers. Demand Flow manufacturing uses blanket purchase contracts with releases against these contracts. Information systems in Demand Flow manufacturing track contracts, transportation networks, and packaging considerations from the purchasing standpoint, as opposed to the traditional use of detailed scheduled purchase orders. On average, Demand Flow manufacturing requires 60 percent fewer trans-actions than MRP II.

System Thoughts

Information systems can be a powerful tool in the conversion from schedulized subassembly manufacturing to Demand Flow manufacturing.

MRP II systems employ standard routings with no relationship to the TQC sequence of events. The traditional MRP II routing system relies on the bill of material to structure independent processes, fabricated parts, and subassemblies. In Demand Flow manufacturing, the bill of material is the equivalent of a "pile of parts," and the TQC sequence of events controls the process. The TQC sequence of events requires information-system support to identify which events of the sequence of events are value added and non-value added, which steps are setup, which events are move events, and, most important, what are the quality criteria for each element of work. Once the relevant information is in the system, the system can help identify the targeted work content and quality criteria for each operation.

This identification process is based on the targeted operational cycle-time calculation established during line design. This is a very valuable tool during the initial flow line design, as well as in the ongoing process-improvement design changes. The system can also assist in the identification of any non-value-added events in the TQC sequence of events. It becomes a valuable management tool in identifying the problem areas to be addressed when reducing the total product cycle time.

The bill of material system in Demand Flow manufacturing must include the backflush location information and, in many cases, intermediate deduct-identification information. The Demand Flow manufacturing bill-of-material computer system should also include a "pending file" in the engineering-change system. The "pending file" is essential in maintaining the correct backflush effective dates. It authorizes the procurement of new materials based on the approved date of an engineering change, but it does not change the actual bill of material in the process until the part(s) are obtained and consumed in production, and the product has reached the point of completion and backflush.

Flexible Demand-Based Planning Systems

The Demand Flow manufacturing rate-based planning system must offer functions that typical MRP II scheduling systems do not perform. This includes such features as flex-fence management tools, forecast-consumption tools (to blend forecasts and actual orders), the ability to plan and view resource requirements (without using work orders, daily flow rates, and resource calculations to see how many people it takes to support a particular daily rate), total product cycle time calculations (to recommend demand and planning time fences), daily forecast-violation reporting, actual flow-rate performance, and linearity reporting in the production process. The daily rates are established at the back of the process as opposed to the scheduling techniques and systems that attempt to control the front of the process.

Formal "Pile of Parts" and Engineering Tools

The bill of material is important in traditional manufacturing, but it is doubly critical in Demand Flow manufacturing. It not only controls the parts to buy, but it also controls inventory. The bill of material functionality of a Demand Flow system should include the capability to compress a traditional multilevel bill of material into a single level bill of material. The design of a Demand Flow system should give the user the capability to determine whether the subassembly part number should be eliminated, restructured independently as FRU, or remain as an additional level on the bill of material.

Other major changes in the bill of material's format include the Demand Flow manufacturing "pending file." This allows the material-requirements-planning algorithm to correctly identify and plan the purchase of parts associated with the engineering-change order based on the approval date of the engineering-change order, but it does not yet modify the bill of material for backflush or configuration-control purposes. This means that the revision-level logic of many systems must change so that the revision level is not applied until the actual engineering change has been incorporated in the product through production. The bill-of-material system must also contain backflush locations and deduct-identification information.

Although Demand Flow uses only one bill of material, different lines (line IDs) in different plants can have different backflush locations and deduct-identification information. The bill-of-material "where used" system must have the capability of identifying which TQC operational sheet contains a specific part number. If an engineering change affects a specific part number, the design-and-process engineer needs to know what sheets are affected when the change is approved.

Backflush and Kanban Inventory Control

The bill of material also must identify the backflush information required for inventory management and control. Intermediate backflush capability at a user-defined deduct point also should be available. The single-level bill of material is key to lowering the number of transactions. In traditional manufacturing, material-requirements planning individually processes through each level of the bill of material, regardless of any requirements for each level of the bill of material. The information system processes those levels one by one. This is one reason material-requirements-planning runs in large companies may take an extended period of time to complete and process. As a result, many companies do not execute this planning process each day. Because of the Demand Flow manufacturing flat bill of material—along with the elimination of the work-order-system logic—material requirements planning can now run in a fraction of the time.

TQC Operational-Sheet Coordination

The bill of material also should include TQC operational-sheet information, and the sheet identification number should be tied to a line item on the bill of material. As discussed earlier, when an engineering change is made to the flow-manufacturing bill of material, the system can identify the affected sheets and those that may need review and modification. An eventual goal is to link the manufacturing system bill of material with the CAD design systems. With that connection to the design process, the manufacturing system can directly use CAD information to aid in TQC operational-sheet design and bill-of-material creation. Although separate organizations may manage the bill-of-material information, one bill of material should exist.

The design or product-engineering group controls the pile of parts—whichever group is ultimately responsible for the form, fit, and function of the product. Planning or production controls the backflush location and deduct-identification information. The security for the change capability of the bill of material must be segregated accordingly.

Critical Engineering Information

The bill-of-material system is where all of the critical engineering change and revision information resides. If a revision level change is made to a part number, the old revision level, new revision, and the nature of the change with the engineering change number should be retained by date. If an engineering change requires a part-number change, the ECO (engineering change order) "pending file" is used until the change is actually incorporated into the product. Two types of change information exist: current and pending engineering-change information. The pending bill of material, based on its approval, guides the procurement of material, while the current bill of material assists in production and the backflush material out of raw in-process (RIP) inventory. Once the change is physically implemented, the status of the changed bill of material is moved from pending to current.

Kanban Management

The daily management requirements of a kanban system should not be underestimated. Systems that support Demand Flow manufacturing are invaluable in identifying kanbans that are in RIP, along with the information related to pull sequence and quantity. In addition to managing the current kanban, systems that support Demand Flow manufacturing must facilitate the creation of new kanbans for new products based on existing lines. kanban management also must identify any obsolete kanbans that are no longer required in current product lines. The sizing, creating, and printing of kanban cards must follow replenishment parameters. Although kanban execution is quite simple, the requirements of effective kanban management should not be underestimated, particularly in high-growth and high-technology companies.

Managing Flow and Flexibility

The Demand Flow production system should assist in calculating operational cycle time, resource requirements, and total product cycle time for each line or cell. This assistance is vital in the design and management of the Demand Flow process.

The demand-based planning system should assist with managing the master-planning information and the conversion of this data into a daily rate. Based on the manufacturing calendar, a formalized Demand Flow system should track actual linearity against the planned daily rate. For purchasing, the system should track the negotiated flexibility contained in the purchase contracts and should help determine the frequency of parts delivery based on an algorithm that balances transportation costs against inventory carrying costs. The system also should monitor DFT contracts and measure supplier performance on both early and late deliveries. Delivery performance, flex-window management, and quality-performance information are vital to the management of the DFT company.

Activity-Based Accounting

A Demand Flow cost-accounting system is set up for activity-based accounting—applying overhead to a process and using total product cycle time as the key to applying the overhead. Demand Flow manufacturing requires a change from labor-based to activity-based accounting systems. Labor now becomes an element of overhead. The Demand Flow process includes only two costing elements: material and overhead. Overhead includes a fixed percentage (applied across all lines in the plant) and a variable overhead rate (applied at specific operations or cells within each line). A DFT system should be able to track process variance, with an emphasis on the total product cycle-time goal versus actual and product-costing variance (based on the standard overhead and material credited versus the actual overhead required and actual material consumed).

Inventory Balances Are Backflushed

Systems that support Demand Flow manufacturing will need to track the inventory balance in RIP, as well as the inventory balances in the storerooms. The inventory-control system in the storeroom probably keeps the same location-tracking system used in traditional manufacturing as long as a storeroom is required. RIP inventory is categorized independently as a general-inventory location. RIP inventory has no relationship to any particular location. The formal system sums up the inventories in RIP and in the storeroom for a total inventory balance used for planning purposes.

Computers Reflect Processes

The differences between a Demand Flow system and traditional MRP II are obvious when comparing processes. The Demand Flow manufacturing pull system uses the following:

- A demand-based system for planning long-range material requirements
- Releases generated against a blanket purchase order for the preferred single supplier
- Material receipts directed to RIP inventory
- Materials relieved by a backflush transaction

This is in direct contrast to traditional MRP II, which uses schedules in a push fashion to control the schedule of purchase orders and work orders. After scheduling the order, the system issues material from a storeroom and releases the work order to production. Purchased parts are received from a particular purchase order line item and transacted into the storeroom until required for issue to a work order.

Graphic Orientation

The computer tools and techniques employed in Demand Flow Technology are very graphically oriented. The system uses many charts and graphs in reporting data for analysis. Graphic techniques also help create pictorial TQC operational sheets, charts on total-quality performance, improvement Pareto charts, process capability analyses, fishbone diagrams, and so forth. From a manufacturing engineering standpoint, advanced graphics techniques help create graphic-production documentation, as opposed to the traditional text-oriented production documentation.

These graphic production documents, or TQC operational sheets, represent an exceptional tool for the quality and performance of work in the process. Production employees can now perform verification and total quality control. These sheets feature a large color illustration that graphically directs the operator to points of work and verification. Red lines or other manual modifications are no longer tolerated, as they defeat the TQC thrust of the operational-method sheets. Personal computers and technical-illustration software can be used to create such sheets in minutes.

Transition Strategy

Fortunately, the priority of making the transition to systems that support Demand Flow Technology is secondary to the actual implementation of the methodology itself. This does not understate the importance of the supporting information system. Comprehensive simplicity and the reduction in transactions and associated overhead are key to success. Better integration with suppliers' material management is the natural outcome of improved partnerships—providing benefits that support the speed of execution within the flow lines.

Chapter 15
Embracing the Strategy
and Implementing the Technology

World-class Demand Flow manufacturing is a company-wide business strategy. It recognizes that manufacturing can provide a distinct advantage for competing in the global, on-demand economy. This competitive advantage comes from Demand Flow© Technology manufacturing and its focus on speed to market and market responsiveness. Demand Flow manufacturing technology reduces new-product-development time and substantially increases speed to market. Demand Flow manufacturing seeks to highlight and then eliminate unnecessary work—in other words, anything that does not add value in the eyes of the customer.

Flexible people, employee involvement, TQC operational sheets, kanban management style, TQC product development, and flow-line designs are keys to the market responsiveness in Demand Flow manufacturing. DFT's objectives include world-class quality and high inventory turns, which lead to a higher return on assets.

Technology turnover—the rate of new-product introductions—used to be 5 to 10 years. Today, it is 12 to 24 months. Demand Flow manufacturing technology not only sets the stage for manufacturing excellence, it is essential to the survival of a manufacturing company. Demand Flow Technology depends on flexible people, customer satisfaction, demand-based manufacturing, and TQC-process technology.

Organization for Implementation

The company-wide project team takes shape at the beginning of the implementation process. A full-time project leader is named. This individual reports to the division manager, typically the vice president of manufacturing or plant manager. The team leader should have a solid and broad familiarity with the business, be well rounded in the technology and strategy of Demand Flow manufacturing, and be able to get things done despite the inevitable obstacles. The leader must have strong interpersonal skills, because people are essential to the successful DFT implementation. An individual with experience in crossing organizational barriers and overcoming entrenched mindsets is mandatory. The leader *must* be a current employee in the division where the implementation is taking place (see Figure 15.1).

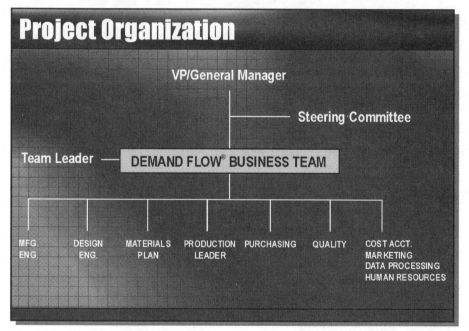

Figure 15.1 — Project Organization

Single-Level Management Team

The team leader is the cornerstone of the implementation steering team. This steering team "owns" the transformation program and must be chaired by the senior management "champion," most often the manufacturing vice president or plant manager. The DFT project leader is a member of the steering team and selects an implementation team to manage the transition. Once selected, this implementation team spends a large amount of time on the day-to-day activities of DFT implementation. Although they maintain their regular positions, the DFT project is their primary responsibility. The team consists of representatives from design engineering, finance, manufacturing engineering, manufacturing systems, quality, production, planning, and materials. They form the core group for implementation of the specific plant or product line. Support members from human resources, sales, CAD/document control, and information technology are also on the team, but they act in an advisory capacity. When the implementation reaches a major milestone, the core implementation team members, with the exception of the team leader, often rotate back into their functional positions, and new team members are selected. This approach adds specific product and process expertise to the implementation, as well as providing many people with the opportunity to participate actively in the implementation process. The implementation steering team must be hands-on individuals with the authority and responsibility for implementation. The steering team approves all techniques and methods before the major implementation tasks even begin.

Implementation Starts from the Back

The company now selects a plant and product for DFT implementation. The implementation should always begin at the back stages of production or at final assembly. Feeder lines and upstream operations can continue to batch and schedule, but daily rates of quality units are initially targeted at the end of the process. An area that is important to the business and has high visibility should be selected. These implementations must not fail—and will not fail—if the company maintains its focus on Demand Flow Technology. Also, the business feels the positive impact sooner by selecting a key product or process.

Compromises during the early implementation phase can be extremely dangerous. Early successes must come within three to six months of the implementation kickoff. Any anticipated or emerging problems should not slow or threaten the project. The management steering committee must address them as quickly as possible.

Designed with a Consistent Technology

Any new products still in the development stage should be managed and released in a Demand Flow manufacturing mode. This is much easier and more cost effective than taking a product designed in a traditional manner and trying to adapt it for Demand Flow manufacturing.

Create an Implementation Plan

The DFT implementation should proceed in an orderly and planned fashion. Each step must be well planned and take into consideration all potential obstacles. The company should complete each step before moving to the next. Omitting a key step can lead to negative results and the misperception that Demand Flow manufacturing "can't work here." Implementing little pieces of Demand Flow Technology here and there and keeping business as usual everywhere else is a sure recipe for failure.

Commitment from the Top

The implementation process should begin with the creation of a strategic master-implementation plan that identifies product/process priorities and key goals. Management must support the company-wide, world-class strategy and become familiar with the world-class concepts and technologies. Managers should also continually demonstrate their commitment and involvement. An implementation organization is then formed to define clearly stated objectives, responsibilities, and deadlines. Demand Flow manufacturing implementations should not become career jobs. Positive results should emerge within a targeted time frame.

Educating on One Track

An education plan is also critical. Management, the implementation team, and other key personnel must receive consistent, detailed Demand Flow manufacturing education. There are many types of lead-time-reduction and inventory-reduction methodologies, but there is only one world-class Demand Flow manufacturing technology. All employees must share the same technology. Demand Flow relies on new terminology, and employees across the company must fully understand the new terms and language. Variations in approach can swiftly derail the implementation. The deeper an organization's education in Demand Flow Technology, the greater the chances for success within the planned time frame.

Follow Steps; Make Determinations

The next step is to start the development of the Demand Flow production process. This begins with establishing a TQC sequence of events and a product synchronization for each product. This natural flow of the product identifies work content, TQC points, and non-value-added steps. Then, based on the highest desired manufacturing rate (demand at capacity), a targeted operational cycle time for a process is determined. TQC operational sheets (which clearly show the work content at each operation, TQC, and verification data) are then generated. Bills of material make the transformation into compressed piles of parts with their associated backflush locations. If necessary, intermediate-deduct points are also identified. Mixed-model lines and cells result from process mapping and group-technology techniques. Total product cycle time is calculated and becomes the basis for inventory investment and the application of overhead. Initial conceptual line designs are now sketched out—but only on paper!

Sales Is Key in Flexibility

In the next stage, negotiations with sales determine the limits of flexibility required to compete and support the customers' needs. Sales establishes the forecast, outlines flexibility windows, and takes responsibility for finished goods. Also in this stage, TQC and employee-involvement programs begin. All employees are trained on simple, statistical problem-solving techniques. Employee-involvement teams are also formed, with training provided on the employee-involvement projects and group dynamics. A responsive and flexible pay and reward mechanism should be planned at this point as well. A training program to certify a minimum of "one up, one down" employee flexibility within the process is also established.

During this stage, procurement begins discussions with suppliers. Procurement holds a "supplier day" event to inform suppliers that, while the company hasn't yet implemented Demand Flow manufacturing, the implementation is a strategic goal. The company eventually will work more closely with a smaller number of suppliers, requiring a greater emphasis on flexibility and quality. Suppliers also receive the requisite education on kanban pull strategies.

Plan, Design, and Do It

Based on the line design completed earlier, the next stage involves calculating in-process and RIP material kanban for each operation. Once this is complete, the physical change of the line layout and the conversion of the back portion of the process to a Demand Flow environment can proceed. Once the Demand Flow process achieves 90 to 95 percent linearity, the pull process can expand to include supplier pulls. Transportation networks, DFT supplier contracts for class "A" parts for the target line, and RIP areas and pull sequences should all be ready for production.

At this time, the kanban and inventory management includes the backflush technique based on the Demand Flow manufacturing "pile of parts" bill of material. Also, testing of the costing system to a total product cycle time basis and completing the modification of compensation and organizational systems are now completed.

Management reports produced at the end of the day ensure the daily rates are achieved in a productive and flexible flow process. Minor changes in kanban sizing and operational-sheet management should be expected, as should minor modifications to the previously tested formal system. Properly designed Demand Flow lines should become linear within a week of the physical line change—up to an initial target of 85 percent.

When a manufacturer follows and completes this implementation plan, it has a very simple yet powerful manufacturing advantage that can compete with any company in the world.

Survival or Leadership

The world is changing. The results and techniques that worked in the past are not good enough in a highly competitive, demand-driven market. Tomorrow's leaders are the visionaries who pursue results that are a *Quantum Leap* beyond those achieved with traditional techniques, methods, and systems.

Taking the Right Turn

Many manufacturing industries are at a crossroads. In one direction lies the continually waning competitive stature, an ongoing trend toward a service-based economy, and the inevitable decline in the standard of living for people who would otherwise find employment in manufacturing. The other direction requires a renewed commitment by individuals and companies to be the best in the world. Being the best means being the leader in the speed-to-market implementation of technology—producing the highest-quality products at the lowest possible cost and using manufacturing as a profit-generating competitive advantage.

However, the road to being the best is not a path of gentle evolution. Leaders with vision must abandon outdated manufacturing techniques. They must overcome constraining mindsets, grasp Demand Flow Technology, and make the *Quantum Leap* to *world-class* excellence.

Appendix A
Further Reading

Built to Last: Successful Habits of Visionary Companies. Jim Collins and Jerry I. Porras. Harper Business, 2002.

ERP: Tools, Techniques, and Applications for Integrating the Supply Chain. Carol A. Ptak and Eli Schragenheim. St. Lucie Press, 1999.

5S for Operators: 5 Pillars of the Visual Workplace (For Your Organization!). Hiroyuki Hirano. Productivity Press, Inc., 1996.

Gemba Kaizen: A Commonsense, Low-Cost Approach to Management. Masaaki Imai. McGraw-Hill, 1997.

Good to Great: Why Some Companies Make the Leap...and Others Don't. Jim Collins. Harper Business, 2001.

Kaizen: The Key to Japan's Competitive Success. Masaaki Imai. McGraw-Hill/Irwin, 1986.

Lean Thinking: Banish Waste and Create Wealth in Your Corporation. James P. Womack and Daniel T. Jones. Free Press, 2003.
MRP and Beyond: A Toolbox for Integrating People and Systems. Carol A. Ptak. McGraw-Hill, 1996.

Necessary But Not Sufficient. Eliyahu M. Goldratt, Eli Schragenheim, and Carol A. Ptak. North River Press, 2000.

Orchestrating Success: Improve Control of the Business with Sales & Operations Planning. Richard C. Ling and Walter E. Goddard. John Wiley & Sons, 1995.

The Balanced Scorecard: Translating Strategy into Action. Robert S. Kaplan and David P. Norton. Harvard Business School Press, 1996.

The Goal: A Process of Ongoing Improvement. Eliyahu M. Goldratt and Jeff Cox. North River Press, 1992.

The Machine that Changed the World: The Story of Lean Production. James P. Womack, Daniel T. Jones, and Daniel Roos. Perennial, 1991.

The Strategy-Focused Organization: How Balanced Scorecard Companies Thrive in the New Business Environment. Robert S. Kaplan and David P. Norton. Harvard Business School Press, 2000.

What Is Six Sigma? Peter S. Pande, Lawrence Holpp. McGraw-Hill. 2001.

Appendix B
Useful Websites

Association for Manufacturing Excellence — www.ame.org

American Production and Inventory Control Society — www.apics.org

The Balanced Scorecard Institute, Dr. Robert Kaplan and Dr. David Norton — www.balancedscorecard.org

Dr. Barry Lawrence, Texas A&M University — http://etidweb.tamu.edu/lawrence

International Society of Six Sigma Professionals — www.isssp.net

JCIT International, Demand Flow© Technology — www.jcit.com

The Lean Enterprise Institute, James P. Womack — www.lean.org

The Supply Chain Council — www.supply-chain.org

Supply Chain Systems Laboratory, Texas A&M University — http://supplychain.tamu.edu

Appendix C
Manufacturing Organizations

American Electronics Association — www.aeanet.org

British Plastics Federation — www.bpf.co.uk

CAD Society — www.cadsociety.org

California Cast Metals Association — www.foundryccma.org

Chicago Association of Spring Manufacturers — www.casmi.org

Dangerous Goods Advisory Council — www.hmac.org

Data Interchange Standards Association — www.disa.org

European Organization for Conformity Assessment — www.eotc.be

Independent Lubricant Manufacturers Association — www.ilma.org

Industrial Distribution Association — www.ida-assoc.org

Industrial Supply Manufacturers Association — www.ismaonline.org

Institute for Supply Management — www.ism.ws

International Committee for Information Technology — www.ncits.org

Manufacturing Enterprise Solutions Association — www.mesa.org

Material Handling Equipment Distributors Association — www.mheda.org

Metal Finishing Suppliers' Association — www.mfsa.org

Minerals, Metals, and Materials Society — www.tms.org

National Association of Energy Service Companies — www.naesco.org

National Association of Manufacturers — www.nam.org

National Association of Printing Ink Manufacturers — www.napim.org

National Coalition for Advanced Manufacturing — wwwnacfam.org

National Institute of Standards and Technology — www.nist.gov

Occupational Safety and Health Administration — www.osha.gov

Office of Industrial Technologies — www.oit.doe.gov

Packaging Machinery Manufacturers Institute — www.pmmi.org

PCI Industrial Computer Manufacturers Group — www.picmg.org

Plastics Institute of America — www.plasticsinstitute.org

Quality, Engineering and Manufacturing Association — www.tqm.org

Robotics Industries Association — www.robotics.org

Robotics International, Society of Manufacturing Engineers — www.sme.org/ri

Rubber Manufacturers Association — www.rma.org

Small Motor and Motion Association — www.smma.org

Society for Information Management — www.simnet.org

Technical Association of the Pulp and Paper Industry — www.tappi.org

Textile Rental Services Association of America — www.trsa.org

The Association for Manufacturing Technology — www.amtonline.org

The Supply-Chain Council — www.supply-chain.org

Valve Manufacturers Association of America — www.vma.org

Appendix D
Glossary of Lean and
Flow Manufacturing Terms

ACQUISITION COST
The costs associated with purchasing material—usually expressed as a percentage of the purchased-material standard cost. Acquisition costs typically include procurement, shipping, and inspection costs.

AVAILABLE TO PROMISE (ATP)
The quantity of a product that is available to commit to customers. Available-to-promise quantities exist when total demand exceeds actual orders.

BACKFLUSH
The method in Demand Flow manufacturing of relieving RIP inventory from a product's bill-of-material quantities/usage at the end of the flow process when the product is complete. Backflush also can relieve purchase orders of consigned inventory material.

BALANCED POINT
The point in time at which plant capacity and resources balance exactly with market demand.

BATCH MANUFACTURING
The traditional manufacturing philosophy under which companies schedule production using lots or quantities. Batch manufacturing usually includes work orders created corresponding to a production schedule for a quantity of fabricated parts or subassemblies.

BILL OF MATERIAL
A hierarchical listing of materials, both purchased and manufactured, required to build a product.

BOTTLENECK
A point in the production flow where resource capacity is less than the demand for the resource.

CAPACITY-CONSTRAINED RESOURCES
Resources whose availability is insufficient to meet current or ongoing demand.

CELL
A grouping of dissimilar resources (people and/or machines) positioned in a logical sequence to facilitate the flow of a product or products through the production line. Also see Group Technology.

CHAKU-CHAKU
A method of conducting single-piece flow, where the operator proceeds from machine to machine and takes the part from one machine and loads it into the next.

CONSTRAINT
A process that prevents a system from achieving higher performance or throughput.

COST OF GOODS SOLD (COGS)
The costs related to purchasing raw materials and manufacturing finished goods. Sometimes referred to as "cost of sales."

COVARIANCE
The impact of one variable upon others in the same group.

DAILY RATE (D_R)

The daily production quantity required to meet customer demand for that day. The daily rate determines a line's staffing requirements for the day.

DEMAND AT CAPACITY (D_C)

The highest-targeted-volume output of products a specific plant or line can achieve. Demand at capacity is the management-defined quantity that the Demand Flow process must support, usually stated in terms of products per day. This target is rarely if ever adjusted.

DEMAND-BASED PLANNING

A production and purchased material-planning methodology used in Demand Flow Technology. Demand-based planning exclusively employs demand-time fences, planning-flex fences, and total-demand calculations.

DEMAND-BASED MANAGEMENT

See Demand-Based Planning.

DEMAND-TIME FENCE

The point in time when actual demand and forecasted demand combine to become total demand. Demand cannot be adjusted within the demand-time fence.

DEPENDENT EVENTS

Events that occur only after a previous event.

DUAL-CARD KANBAN

A demand-pull technique that uses a "move" and "produce" communication method. Dual-card kanbans are typically used in machine-intensive manufacturing processes or independent cells with intensive setups and long replenishment times.

ERP

Enterprise requirements planning. ERP is functionally similar to MRP II, with the additional capabilities for computer-aided design (CAD), computer-aided manufacturing (CAM), electronic office communications, and other financial and distribution systems.

EVAPORATING CLOUDS
A method of resolving the conflict between two apparently mutually exclusive alternatives by defining the shared goals and objectives of each alternative.

EXTERNAL SETUP
Die-setup procedures that can be performed while machine is in motion. Also known as OED, or "outer exchange of die."

FRU
A field replaceable unit, or spare. In Demand Flow Technology, an FRU is treated as an independent product with its own forecasting, manufacturing processes, and costing. It is typically produced in a mixed-model line with other complete products.

FEEDER PROCESS
A branch process that feeds directly into a consuming operation or flow process. The feeder is always identified independently from the consuming process on the product synchronization. (See Product Synchronization).

FLOW KAIZEN
"Kaizen" denotes a mindset of continual and gradual improvement resulting from small, incremental improvements. In a flow environment, Kaizen is usually applied only once within a value stream.

FLOW RATE
The current daily rate, divided by effective work hours in a shift, multiplied by the number of shifts in a day. The flow rate represents an important volume-related management criteria that should be adjusted daily, based on current customer demand.

FORECAST CONSUMPTION
A technique used to determine the production plan, initially using a blend of the forecast and sales orders and later based on actual orders alone. Forecast consumption is a demand-based planning tool to coordinate marketing, production, and outside suppliers.

GROUP TECHNOLOGY
The practice of organizing people and different functional machines into cells to produce related parts or products. The focus of the cell is to reduce inventory, reduce or eliminate queues, improve quality, and reduce throughput times.

HEIJUNKA
The practice of keeping total manufacturing volume as constant as possible.

HOMOGENEOUS OVERHEAD
A pool of fixed or variable overhead costs required to convert material into a finished product. Homogeneous-overhead-cost pools typically apply to all products being produced.

HOSHIN KANRI
The process of defining goals, outlining the projects required to achieve those goals, assigning the people and resources needed to complete the projects, and establishing relevant project metrics

INFORMATION-MANAGEMENT TASK
The task of moving a specific product from the initial customer order to detailed production scheduling to delivery.

IN-PROCESS KANBAN
An inventory of component material or in-process inventory required to support designed imbalances among DFT operations. An in-process kanban never has a part number identity and is usually represented by the letter "X" printed on the card, table, or floor.

INTERMEDIATE BACKFLUSH
An inventory transaction performed to relieve RIP inventory that is consumed during production up to a specified physical location in the manufacturing line. Material relieved from RIP is considered to be in a "holding" in-process bucket that will later be relieved through either an end-of-the-line backflush transaction or a scrap transaction. An intermediate backflush can be used as a physical point for scrap transactions, as well as for process-audit-inventory purposes.

INTERMEDIATE-DEDUCT ID
A physical location in a Demand Flow line where an intermediate backflush occurs.

INVENTORY CARRYING COSTS
The actual costs associated with maintaining inventory. These costs apply to raw, RIP, and finished-goods inventory. The cost of carrying inventory is typically 25 to 40 percent of the value of the inventory. Inventory carrying costs may include the cost of money, inventory storage, inventory management, scrap and obsolescence, taxes, and lost opportunity.

INVENTORY TURNS
The number of times inventory turns over in one year. To calculate inventory turns, divide the annual estimated inventory requirements by current total on-hand inventory.

JIT
Just in time. A term associated with a manufacturing technique based on flow and pull manufacturing processes rather than traditional scheduling techniques. JIT is often misinterpreted as an inventory-reduction program. The JIT technique is only a very small component of Demand Flow Technology.

KANBAN
The Japanese word for "communication signal" or "card." kanban is a technique for pulling products and material through and into the Demand Flow manufacturing process. It can be a physical signal, such as a container or card. Information on a raw-material kanban identifies part number and description information, as well as points of usage/consumption and supply/replenishment, along with a calculated quantity.

LABOR HOURS
The time required to complete the required manufacturing steps to meet design or production specifications.

LINEARITY
The relationship of planned daily rates versus actual production, as monitored at the back of the production flow (completion). One of Demand Flow's primary objectives is to adjust the volume and mix of products every day based on actual customer demand and in an effort to meet the planned daily rate.

MACHINE CELL

A grouping of dissimilar machines in a logical sequence to facilitate the flow of a product or products. In a traditional manufacturing environment, the work of machine cells is tightly scheduled. In a Demand Flow manufacturing environment, kanbans pull work and materials through machine cells. See Group Technology.

MACHINE TIME

The time required for machines to complete the manufacturing steps required for products to meet design or product specifications.

MIXED-MODEL DEMAND FLOW LINE

The design of flow lines to produce families of similar products. The mixed-model Demand Flow line has the ability to accommodate a range of volumes for any product, any day, based on the direction of actual customer demand. Quality is designed into the line and executed through TQC operational method sheets.

MIXED-MODEL SEQUENCING

The order in which the total demand for products and options is to be pulled into a mixed-model Demand Flow line. Sequencing is also a balancing technique used to minimize imbalances among the products to be produced in the same mixed-model Demand Flow line.

MOVE TIME

The time spent in moving products or materials from one point to another, either through human or machine labor. Appreciable move time usually indicates a poorly designed line.

MRP

Material requirements planning. The outdated computerized scheduling system used to schedule material requirements from suppliers and internal production.

MRP II

Manufacturing resource planning. A later, also outdated, version of the MRP computer-scheduling system. MRP II systems have additional sophistication to track production through work centers and departments.

NONREPLENISHABLE KANBAN
A type of material kanban that is not replenished when emptied. Manufacturers use nonreplenishable kanbans for custom products, one-time customer orders, and infrequent material demands. Nonreplenishable kanbans require distinct management processes.

NON-VALUE ADDED
Steps in the production process that may be necessary but do not increase the value of a product or service for the customer.

OPERATION
A grouping of tasks (work elements) from an SOE (sequence of events) that defines the work to be performed. The cumulative time for this grouping of tasks should be equal to or less than the targeted operational cycle time. In DFT environments, an operation is work performed by a person or a machine.

OPERATIONAL CYCLE TIME
The calculated target of work content time to be performed independently by a person or machine on a Demand Flow line. To calculate operational-cycle time, multiply the effective work hours by the number of shifts, and divide the result by the designed capacity of products to be produced in the Demand Flow line. Design capacity is specified for each product, and it is determined by management and marketing projections.

OPERATIONAL TQC METHOD SHEETS
A color graphical representation of quality criteria and work content as defined by the sequence of events. TQC method sheets display little or no text and rely on graphics to communicate the work content of an operation as well as the verification steps and requirements to ensure the total quality of a previous operation.

PARETO
Pareto is the act of tracking the occurrences and frequency of opportunities for improvement at every operation in a process.

PARTICIPATORY MANAGEMENT
A departure from traditional departmental management and a move toward team management. Includes membership from multiple disciplines such as marketing, engineering, production, quality, materials, etc.

PILE OF PARTS
Individual components or parts required to manufacture a product. These parts are not structured in a traditional multilevel, subassembly, or fabricated-part method. This becomes a flat listing of purchased parts required to build a product.

PLANNING-FLEX FENCE
The negotiated flexibility of the time frame outside the demand fence. The total demand and the resulting planned daily rate may fluctuate within predetermined product parameters that are based on flexibility ranges negotiated with marketing and customers. Material requirements are negotiated with suppliers, mirroring the flexibility available to marketing and customers.

PLANNING VISIBILITY
The long-range forecast for requirements used internally and with suppliers for planning purposes. Forecasted requirements with flex-window information are shared with all "pull" suppliers.

PROCESS
A DFT grouping of logical and functional manufacturing steps required to convert material into a completed product. The progression of work in a process is defined as a sequence of events.

PROCESS MAPPING
A matrix of processes defined by product synchronizations. Demand Flow manufacturers use process mapping to determine the commonalities among manufacturing processes and among products. The objective of process mapping is to develop families of products that share common processes to determine which products can be produced in the same mixed-model Demand Flow line.

PRODUCT SYNCHRONIZATION
A design-engineering technique used in the implementation of the Demand Flow business strategy. Product synchronization defines the relationships of manufacturing processes required to produce a product, and the relationships among various process phases. It is the basis for product design and the design of the Demand Flow line.

PULL SEQUENCES

A chain of information that identifies points of usage/consumption and supply/replenishment. A pull sequence is the path along which kanban material is replenished through the Demand Flow manufacturing process. Pull sequences comprise "from" and "to" relationships associated with kanban replenishment time frames for kanban sizing.

QUEUE

The waiting time or inventory buildup in the traditional batch-manufacturing or scheduled-manufacturing environment. Queue time never appears in a TQC sequence of events and is never part of operational definition.

QUICK COUNT

A method of replenishing material kanbans that uses approximate replenishment quantities (e.g., handful, fill, cupful, pallet). This method is always used in RIP when no special need to count the material precisely exists.

RIP

Raw in-process inventory. Excluding any material in the storeroom, RIP includes all raw and in-process materials at supply points or close to the production line and feeder processes. RIP is supplied by the storeroom and/or directly by suppliers. Material quantities pulled into RIP are acknowledged by quantity when entering. Material and production movements from any point to another within RIP are not counted or tracked. Once the product is complete and leaves production, the bill of material is relieved from RIP inventory by a backflush transaction. The only other component counting and tracking within RIP is for scrapped material.

SCRAP

Unusable parts or materials. Demand Flow manufacturing and traditional manufacturing both report scrap as it occurs.

SEQUENCE OF EVENTS
The definition of the required work and quality criteria to build a product in a specific production process. The SOE is never the anticipated work content. It is independent of the production environment, methods, specific machine types, and quality criteria. The SOE represents the natural sequence of events required to produce a product compliant with product- and process-design specifications. It identifies sequential work content and specific TQC information against each task. The SOE is one of the essential techniques of Demand Flow Technology, driving operational definition and acting as the basis for all work, line design, mixed-model planning, and TQC method-sheet design.

SETUP TIME
The required non-value-added steps/work that precede the steps that add value to a part or product. Examples of setup time include unwrapping parts and making tooling adjustments on a large machine.

SINGLE-SOURCE SUPPLIER
A supplier specifically selected over other qualified suppliers to receive all of the expected orders for a particular part or component.

SOCIO PRODUCTION
A relatively new management philosophy in which employees participate extensively in the hiring and decision making critical to company operations. Despite its theoretical appeal, socio production tends to extend the decision-making process and reduces the speed of new product introductions.

SOLE-SOURCE SUPPLIER
The only known supplier of a particular material or component. Manufacturers that require the specific part or material sold by this supplier can only purchase this item from one supplier.

SUBASSEMBLIES
Components that are assembled or processed together to meet a specific design drawing or function. Subassemblies have their own part numbers and bills of material and are used in the final assembly of a finished product.

TAKT TIME
"Takt" is the German word for "rhythm" or "beat." In Demand Flow Technology, takt defines the targeted work content for people and machines to meet the production capacity that the Demand Flow line was designed to achieve. Each process on the dedicated product synchronization or mixed-model-process map may have a different takt time if the products have different volumes or yields. takt time is always defined by the operational cycle time calculation.

TEAM PASS
The tagging of a portion of a manufactured product found to be defective due to faulty workmanship. A team-pass product has passed through two operations where defective work was performed but incorrectly verified and improperly validated using a TQC specification. A team-pass product does not factor into the analysis of the daily-linearity index.

TOTAL DEMAND
The projected (and eventually the planned) quantity of products to be produced. Based on the marketing-flex windows, total demand is independent of the demand for purchased material. Outside the demand-planning fence, total demand is determined by taking the greater of forecast or actual orders. Within the demand-time fence, total demand is the actual customer demand that has been averaged over time. Total demand eventually becomes the daily rate at the back of the production line.

TOTAL PRODUCT CYCLE TIME
The accumulated work content time along the longest path of the product synchronization (TP c/t), beginning with the completion of the product (FGI). Total product cycle is typically less than the total work content hours required to build a product. It can only equal the total time to build a product in a simple product with feeder processes or concurrent work. It also denotes the time inventory is required to support the processes to build a product.

TQC
Total quality control. The technique in Demand Flow manufacturing that enforces quality in the manufacturing process at the point where work is being performed. TQC is defined by the sequence of events and occurs at every step in the production process (in contrast to external inspection points characteristic of traditional manufacturing environments).

VALUE ADDED

Steps in the production process that increase the worth of a product or service to an external customer or consumer. Value-added steps represent product characteristics and process specifications the customer expects and is willing to pay to receive.

WORK CONTENT

The work required to be performed to build a product that complies with specifications. Work content includes setup, move, non-value-added steps, value-added steps, and quality work.

Index